Naturalists' Handbooks 7

Dragonflies

Other books in the series
1 B. N. K. Davis
  Insects on nettles
2 Valerie K. Brown
  Grasshoppers
3 Peter F. Yeo & Sarah A. Corbet
  Solitary wasps
4 Margaret Redfern
  Insects and thistles
5 Francis S. Gilbert
  Hoverflies
6 Oliver E. Prŷs-Jones & Sarah A. Corbet
  Bumblebees

Dragonflies

P. L. MILLER
Department of Zoology, University of Oxford

Plates by Rupert Lee
Figures by Sophie Allington
Figures for the key to adults by David Chelmick
Key to larvae by Graham Vick
Key to adults by David Chelmick

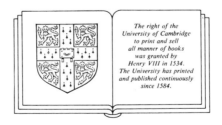

The right of the
University of Cambridge
to print and sell
all manner of books
was granted by
Henry VIII in 1534.
The University has printed
and published continuously
since 1584.

Cambridge University Press
Cambridge
New York   New Rochelle
Melbourne   Sydney

Published by the Press Syndicate of the University of Cambridge
The Pitt Building, Trumpington Street, Cambridge CB2 1RP
32 East 57th Street, New York, NY 10022, USA
10 Stamford Road, Oakleigh, Melbourne 3166, Australia

First published 1987

Printed in Great Britain by Belmont Press, Northampton

British Library cataloguing in publication data

Miller, P. L.
    Dragonflies. – (Naturalists' handbooks; 7)
    1. Dragonflies – Great Britain   2. Insects – Great Britain
    I. Title   II. Series
    595.7'33'0941   QL520.24.G7

Library of Congress cataloguing in publication data

Miller, P. L.
    Dragonflies.
    (Naturalists' handbooks; 7)
    Bibliography
    Includes index.
    1. Dragonflies – Great Britain.   2. Insects – Great Britain.
    I. Title.   II. Series.
    QL520.24.G7M55 1987   595.7'33'0941   86-31719

ISBN 30162 9 hard covers
ISBN 31765 7 paperback

PN

Contents

Editors' preface	*page* vii	
Acknowledgements	viii	
1	**Introduction**	1
	Evolution	1
	The aesthetic appeal and economic importance of dragonflies	2
	The British fauna	3
	Dragonflies are very good animals for field work	4
2	**Eggs and larvae**	6
	Eggs	6
	Larvae	7
	Habitat specificity	8
	Larval feeding	9
	Territorial behaviour in larvae	11
	Larval respiration	12
3	**The adult**	14
	Emergence	14
	Exuviae	15
	Flight	16
	Hovering	19
	Flight metabolism	19
	Thermoregulation	20
	Vision	24
	Feeding behaviour	27
	Maturation	29
	Migrations	30
	Predator avoidance	31
	Reproduction	32
	The structure of the secondary genitalia	34
	The structure of the female reproductive system	35
	Territorial behaviour	35
	Recognition	39
	Intraspecific visual signals	40
	Courtship	41
	Copulation and sperm competition	42
	Guarding	46
	Oviposition	47
4	**Keys to larvae and adults**	49
	Key to larvae by Graham Vick	49
	Key to adults by David Chelmick	57
5	**Conservation and the dragonfly recording scheme**	68
	Dragonfly recording scheme	69

6   Some techniques for studying dragonflies   70
    How to present your findings   72
    Useful addresses   73

    Further reading   74

    Appendix 1. Check list of British species, with English
    names and notes on distribution   77

    Appendix 2. The times of appearance of adult British
    dragonflies   82

    Index   83

Plates 1–4 are between pp. 40 and 41

Editors' preface

Sixth formers and others without a university training in biology may have the opportunity and inclination to study local natural history but lack the knowledge to do so in a confident and productive way. The books in this series offer them the information and ideas needed to plan an investigation, and the practical guidance needed to carry it out. They draw attention to regions on the frontiers of current knowledge where amateur studies have much to offer. We hope the readers will derive as much satisfaction from their biological explorations as we have done.

The keys are an important feature of the books. We are very glad to be able to include in this book a key to adult British dragonflies by David Chelmick and a key to larvae by Graham Vick. Ease of identification also depends on the illustrations. We thank the Natural Environment Research Council for a grant towards the illustrations for this book.

S.A.C.
R.H.L.D.

Acknowledgements

I am very grateful to Philip Corbet, Sally Corbet, Ola Fincke, Laurie Friday, Kate Miller and Michael Siva-Jothy for their numerous useful comments and helpful suggestions.

P.L.M.

1 Introduction

Dragonflies, by which I mean all members of the Odonata including damselflies, comprise one of the most ancient, beautiful and fascinating orders of insects. It is the intention of this book to stimulate interest in and the enjoyment of dragonflies. To many non-biologists the complexity of insect behaviour comes as a great surprise; dragonfly behaviour is no exception and it will continue to provide the naturalist with many opportunities for making original observations in the field and laboratory for many years to come, provided there are dragonflies left to observe. In some regions of Britain, however, over 90% of the aquatic habitats of dragonflies have been destroyed during the present century by agriculture or industry. Only by coming to know and appreciate what we so wantonly and callously destroy can we hope to stop the destruction and preserve what is left. This book is therefore dedicated to the future of dragonflies in Britain.

Evolution

Together with the mayflies (Ephemeroptera), dragonflies make up the Palaeoptera, a primitive group of insects characterised by aquatic larvae and the possession of wings with complex venation and no wing-folding hinge. The Palaeoptera probably evolved in warmer regions about 250–300 million years ago during the Carboniferous, and by the Jurassic many modern families of dragonflies had already appeared. The Carboniferous is known for its giant dragonflies, the Meganeuridae, which belonged to an ancestral group, the Protoanisoptera. Apart from their enormous size, the Meganeuridae differed from modern dragonflies in several ways: for example they lacked certain features such as a nodus and pterostigma (see p. 16) in the wings. Some of the largest fossils, with wing spans of 70 centimetres, come from Commentry in France, but a 50 centimetre specimen was recently found at Bolsover in Derby (Whalley, 1980).*
These must have been impressive giants and their existence poses interesting questions about how they flew, how they avoided overheating, and what they ate.

Modern dragonflies belong to one of three sub-orders:

1. The Anisoptera, which are probably descended from

* References cited under the author's name in the text appear in full in the reference list on p. 74.

the Carboniferous Protoanisoptera, and which have dissimilar fore and hind wings (fig. 8). These are large, robust and fast-flying dragonflies.

2. The Zygoptera, whose antecedents, the Protozygoptera, are known from Permian deposits; in these the fore and hind wings are similar in shape. They are smaller and more slender insects which fly less rapidly than anisopterans and are commonly known as damselflies.

3. The Anisozygoptera, a relict group represented by only two living species, both found in remote streams in the Far East, and both now on the list of endangered species.

The Anisoptera and Zygoptera contain many species which are common in Britain and are familiar sights near fresh water.

Today there are about 5000 living species of dragonflies. They are most abundant in the tropics, occur in a great variety of habitats from semi-desert regions to humid forests, and extend to alpine regions and to the Arctic. Almost all species have aquatic larvae, although in a few tropical and Australian species the larvae have become secondarily terrestrial. In the Petaluridae, an ancient family containing only 9 or 10 species of large dragonflies found in the uplands of New Zealand, Australia, Japan, and North and South America, the larvae are semi-terrestrial, emerging from their water-filled burrows at night to search for prey among moorland vegetation (Winstanley, 1982).

The aesthetic appeal and economic importance of dragonflies

The bright colours of dragonflies have attracted artists for many centuries, and dragonflies have been depicted not only in many paintings, but also in medieval manuscripts, floor tiles, stained glass, and even in postage stamps. They appear in Japanese bronzes as early as 300 BC and they are used today as nose ornaments in Colombia (Asahina, 1974; Geijskes, 1975), while in Malagasy (Madagascar), Indonesia and Malaya they are relished for the table and sold for medicinal purposes (Bodenheimer, 1951). They have also featured in many poems, songs, plays and stories.

Whereas members of the family Calopterygidae (which means beautiful-winged) retain their bright metallic colours after death (pl. 2.1), most dragonflies turn a disappointing brown in the cabinet and they become very brittle. They have not therefore attracted collectors as much as have some other insect orders, and in consequence their natural history has been less well known until recently. Dragonflies are essentially creatures of movement, and their superb aerobatics can be neither guessed at from dead specimens nor captured in photographic snapshots. Colour photography is, however, a very satisfying way of 'collecting' them.

Their economic importance to mankind is not great, although in some habitats their larvae provide important items in the diet of fish, and also prey on the larvae of such carriers of human disease as mosquitoes and blackflies *Simulium*. Likewise the adults sometimes take numbers of tsetse flies, horseflies and mosquitoes, as well as a variety of agricultural pests ranging from aphids to locusts. However, dragonflies are opportunistic feeders, taking whatever prey is abundant; they may reduce the numbers of a pest but are unlikely to eradicate it.

The British fauna

Britain has a small dragonfly fauna of about 45 species (Moore, 1976). In comparison Holland has 69 species (Geijskes & van Tol, 1983) and France about 99, while North America has 332 species of Anisoptera alone.

A brief synopsis of the British fauna is given below.

Sub-order	Family	No. of genera	No. of species
Zygoptera	Platycnemididae	1	1
	Coenagrionidae	6	13
	Lestidae	1	2
	Calopterygidae	1	2
Anisoptera	Gomphidae	1	1
	Aeshnidae	3	8
	Cordulegasteridae	1	1
	Corduliidae	3	4
	Libellulidae	4	13
			45

Of the 45 British species, three are migrants which do not regularly breed in Britain (*Sympetrum flaveolum*, *S. fonscolombei* and *S. vulgatum*), and a further three have not been seen for many years and are probably extinct in Britain, although still occurring on the continent of Europe (*Coenagrion scitulum*, *C. armatum* and *Oxygastra curtisii*) (Chelmick, 1980; Hammond, 1983). New colonies of *Coenagrion lunulatum* were found in Eire in 1982.

Thus there are now 39 breeding species and of these 11 must be considered vulnerable since they have very restricted breeding sites. These are:

Brachytron pratense
Aeshna isosceles   (pl. 1.1)
A. caerulea
Libellula fulva   (pl. 1.4)
Somatochlora arctica
Sympetrum sanguineum
Ceriagrion tenellum
Coenagrion pulchellum
Coenagrion hastulatum
C. mercuriale   (pl. 1.6)
Lestes dryas

Lestes dryas was thought to be extinct in England, but two new colonies were found in Essex in 1983. It is still found in a few localities in Eire.

This means that 28% of the British dragonfly species are threatened, and many of the rest have suffered a severe decline during the last two or three decades as a result of the destruction of habitats by agricultural and urban development and land drainage, and by the pollution of habitats with herbicides, pesticides, excess fertiliser and industrial waste.

The British fauna contains three main elements: Ice Age relics, such as *Aeshna caerulea* and *Somatochlora arctica*; species with widespread distributions all around the north temperate zone, such as *Sympetrum danae* and *Libellula quadrimaculata* (pl. 4.3); and finally Mediterranean species which have moved north as the climate has become warmer, such as *Sympetrum striolatum* (pl. 4.4), *Aeshna grandis* and *Anax imperator* (pl. 3.1, 3.2). The last species has an extensive distribution and it occurs from south Sweden to the Cape of Good Hope. Several of our species spread across Asia to Japan (sometimes as a different subspecies) (e.g. *Lestes sponsa* (pl. 2.5), *Cordulia aenea* (pl. 4.5), *Calopteryx virgo* and *Aeshna mixta*) while others are found commonly also in North America (e.g. *Libellula quadrimaculata* and *Enallagma cyathigerum*) (Corbet and others, 1960).

Notes on the distribution of each species are given in Appendix 1 (p. 77).

Dragonflies are very good animals for field work

Dragonfly larvae are easily kept in aquaria and their predatory behaviour provides a good subject for study (see chapter 2). They can be marked individually and their activity in relation to site selection, defence of territories and prey choice can be examined. There are also fascinating aspects of their respiration and of their emergence to be studied. In the field the distribution and abundance of their different larval stages can form the basis of many projects, and they are important as indicators of pollution, to which some species are extremely sensitive.

Adult dragonflies have been termed bird-watchers' insects because they can be individually marked and watched for many days (chapter 6). They have very interesting reproductive behaviour in which males may defend territories, court females and guard them during oviposition. This can easily be watched in the field by the patient observer. Various attempts have been made to keep dragonflies in captivity, but even by using large outdoor cages this is never entirely satisfactory and most observation is best carried out on free insects. It soon becomes possible for the beginner to make original observations: dragonfly behaviour shows much variability and there is always something new to be seen even in the commonest species. In the following chapters attention will be drawn to several projects which can be done in the field or in the laboratory with simple apparatus.

oviposition: egg laying

The most useful and comprehensive book on dragonflies is that by Corbet (1962). For identification and field notes you should consult Hammond (1983), an excellent book with coloured illustrations, or McGeeney (1986). Useful guides to certain areas have been produced by Butler (1982) for Shropshire, Coker & Fox (1985) for West Wales, Campbell (1983) for Oxfordshire, Dunn (1984) for Derbyshire, and Welstead & Welstead (1984) for the New Forest. A new book on the dragonflies of Europe has been prepared by d'Aguilar, Dommanget & Préchac (1986) and an English translation is available.

The recently formed British Dragonfly Society with over 350 members is extremely active and it organises several excursions and an indoor meeting every year. It publishes a newsletter and a journal, and with the modest subscription of only £4 it offers extremely good value to every dragonfly enthusiast. (See p. 73 for addresses.)

2 Eggs and larvae

Eggs

Most Zygoptera and the Aeshnidae among the Anisoptera insert their long cylindrical eggs into plant tissues. Other Anisoptera, whose eggs are blunter or nearly spherical, deposit them at the water surface or onto plants (fig. 1; see p. 47). One end of the egg bears a microscopic hole, the micropyle, through which sperm enter just before the egg is laid, and the young larva later hatches from this end. The egg shell is smooth in many species but highly sculpted in a few. A jelly is formed by the shell in some species which sticks the eggs to the substrate. In this way the eggs of *Libellula depressa* may be glued to *Ranunculus* (water crowfoot) leaves, or those of gomphids to stones or rocks under the water (Robert, 1958). Some tropical gomphids form long threads on one end of their eggs; these threads may become entangled in vegetation and so prevent the eggs from sinking into soft mud at the bottom or from being washed downstream.

Eggs may be eaten by other animals including fish, which sometimes snap them up as soon as they are laid. They are also subject to parasitism, for example by tiny wasps which may reach them by swimming underwater. Sometimes the eggs start to develop soon after deposition and the larvae hatch 1–3 weeks later. Alternatively, if the eggs are laid in autumn, they may not complete development until the following spring. Some lestid, aeshnid and *Sympetrum* eggs, for example, delay development in this way (Pritchard, 1982). With much care and dedication Robert (1958) has described (in French) the development

Fig. 1. Dragonfly eggs. (*a*) Rounded egg of libellulids, about 0.5 mm long. (*b*) Elongate egg of Zygoptera and Aeshnidae, about 1 mm long.

(a)

(b)

Fig. 2. Patterns of oviposition in plants. (*a*) *Calopteryx virgo*, (*b*) *Platycnemis pennipes*, (*c*) *Pyrrhosoma nymphula*. (After Robert, 1958.)

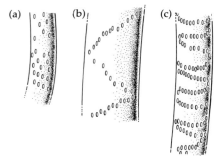

of eggs and larvae in many European dragonflies and his splendid book should be consulted for details; even without a knowledge of French its beautiful illustrations can be enjoyed.

Egg and larval development can easily be examined in the laboratory if plants into which oviposition has been observed are collected and kept in aquaria. Eggs can be searched for by cutting open plant tissue, and the pattern of oviposition can be noted (fig. 2). Alternatively eggs can be collected from females, as described on p. 48.

Larvae

Fig. 3. (*a*) Larva of a zygopteran (*Calopteryx*). (*b*) Larva of an anisopteran (*Orthetrum*).

(a)

(b)

Most of a dragonfly's life is spent in the larval stage and it is among larvae that the greatest range of form is found (fig. 3). Some species have variable numbers of larval moults depending on food supply, temperature and other factors. In *Lestes* there are 6–10 moults, in *Ischnura* 9–12, while some other Zygoptera may have up to 15. Aeshnids commonly have 10–13, whereas libellulids typically have fewer. More details can be found in Lucas (1930) and in Robert (1958).

Development commonly takes 1–2 years, but it can last for as long as 6 years in petalurids and 5 years in some gomphids. Its duration depends partly on altitude and latitude: thus the time *Aeshna juncea* spends as a larva is 4–6 years in north Sweden but only 3 years in central Europe. Other temperate aeshnids complete development in 1–2 years, and in *Anax imperator* in Britain there may be a divided emergence, a few adults emerging after 1 year but most after 2 (Corbet, 1957). In contrast the very large African species, *Anax tristis*, can complete development in 70 days, while some libellulid inhabitants of temporary pools in warm regions do so in as little as 30–40 days. *Ischnura elegans* normally has one generation a year in southern Britain but in the north larvae may live for 2 years; in southern Europe, in contrast, there can be two or three generations in a year.

Rates of larval development depend partly on inherited mechanisms and partly on environmental factors such as temperature and food abundance. The shortening days of autumn can arrest development in some species which then remain without growing throughout the winter. However, by keeping larvae in aquaria at high temperatures, winter emergences can sometimes be induced. Premature development in the winter has been reported as a result of thermal pollution from power stations. Food shortage in the wild is thought seldom to contribute to larval death, but it may cause the developmental period to be extended to a second year in such species as *Coen-*

agrion puella. On the other hand predation on larvae is probably a major factor controlling their abundance. In some habitats fish are the principal predators, but in others water bugs such as *Notonecta*, *Nepa* and *Naucoris* are of great importance and even birds and newts may take their toll.

Habitat specificity

Although a lot of information is available about the types of habitat where larvae are found, there is a need for more detailed knowledge about the microhabitats of different species, the extent to which later instars change their habitats, and the diurnal and seasonal migrations they may undertake. Careful examination of microhabitats at different times of the day and in different months of the year could add much useful new information. Captive larvae can also be offered a choice of conditions in large aquaria – for example by providing different substrates, several types of vegetation or by establishing a temperature gradient by warming one end with an electric bulb outside the tank.

Factors which affect the distribution of larvae may include the pH of the water, the amount and type of aquatic vegetation and whether the water is stationary or running. In Britain acid bogs and peaty areas with sphagnum moss are likely to contain larvae of *Libellula quadrimaculata*, *Pyrrhosoma nymphula*, *Aeshna juncea* and *Sympetrum danae* and, more rarely, *Leucorrhinia dubia* and *Ceriagrion tenellum*, while in the north, *Aeshna caerulea* may also be found. *Cordulegaster boltonii*, *Orthetrum coerulescens* and *Calopteryx virgo* prefer small streams whereas *Calopteryx splendens* and the rare *Libellula fulva* may be found in slower and muddy (but unpolluted) rivers. Small ponds may contain *Libellula depressa*, *Aeshna cyanea* and *Sympetrum striolatum*, whereas larger ponds and lakes may also have *A. mixta*, *A. grandis* and *Orthetrum cancellatum*. Temperature is another important determinant of distribution: *Ischnura pumilio* and *Coenagrion mercuriale*, for example, do not tolerate a minimum February value below about 2 °C, whereas the northern *C. hastulatum* seems to need a minimum of about 0.5 °C (Chelmick, 1980).

Some species tolerate a wide range of conditions and can invade new types of habitat. For example *Platycnemis pennipes* is normally found in rivers, but it also breeds successfully in a few lakes; *Calopteryx splendens* sometimes breeds in canals and ponds; gravel pits in southern England have attracted *Aeshna grandis*, *A. mixta*, *Orthetrum cancellatum*, *Erythromma najas* and *Enallagma cyathigerum*,

instar: a larval stage between two moults

although these all breed in other types of habitat, the last even in moorland tarns. The larvae of *Aeshna mixta* and *Ischnura elegans* can tolerate high levels of chloride and may be found in salt marshes (Kiauta, 1965); indeed *I. elegans* is one of the most widespread and tolerant species, often being the last to disappear from increasingly polluted waters.

In recent years *Anax imperator* and *Aeshna cyanea* have spread further north in England while *A. mixta* and *Enallagma cyathigerum* have become generally more abundant. Many other species, however, have declined drastically. Although such changes may sometimes reflect long-term climatic trends, they are more usually caused by changes in agricultural practice or by gravel extraction and pollution, to which we return in chapter 5. Much more information on larval habitats can be found in Corbet and others (1960) and in Hammond (1983).

Larval feeding

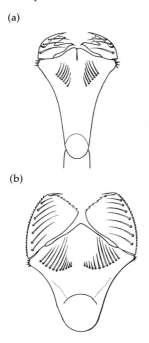

Fig. 4. The larval labial mask. (*a*) A zygopteran, (*b*) an anisopteran.

(a)

(b)

Dragonfly larvae possess a highly specialised mouthpart, the labial mask, which can be shot out rapidly, grasping small prey animals with the hooks at the tip (Pritchard, 1965) (fig. 4). Tanaka & Hisada (1980) have shown that the strike mechanism in *Aeshna* depends on energy stored by muscles and elastic structures in both the mask and the abdomen. About half a second before a strike the valves of the anus close and the abdominal muscles build up pressure equivalent to about 120 centimetres (cm) water. Shortly before the strike the muscles that retract the mask, and also those that extend it, contract together, holding the mask against the face and storing energy. The strike is initiated by relaxation of the mask retractor muscles. It is completed in 16–25 milliseconds and it depends on the sudden release of the stored elastic energy. It therefore allows the predator to strike at a rate almost unaffected by temperature, whereas on cold days a prey's escape response may be sluggish. It is interesting to note that the jump of grasshoppers also depends on stored elastic energy and that it too is relatively unaffected by temperature.

Dragonfly larvae detect prey by sight, by touch, or by both means. After the second or third moult, larval *Orthetrum* and *Libellula* live on the bottom, usually buried, and they find such animals as chironomid midge larvae or oligochaete worms through the tactile responses of their legs and antennae. Gomphid larvae bury themselves in the sand of rivers, and *Cordulegaster* in the beds of streams, behaving similarly to *Orthetrum* and *Libellula*. Corduliids have particularly long legs which increase the area over which they can respond to the vibrations of prey. When a stimulus is detected by such a sprawler or burrower, the larva turns towards it and points the antennae in the direction of the stimulus. When in range it then releases the strike.

The larvae of *Calopteryx splendens* and *Cordulia aenea* live in vegetation but still depend largely on touch or vibration receptors for prey location. Caillère (1974) showed that when *Calopteryx* was deprived of one antenna it could learn to use the other alone to localise prey. *Calopteryx* may leave the vegetation at night and wander in search of prey on the bottom.

The compound eyes of sprawlers and burrowers develop slowly, but the other main group of larvae, the climbers, depend more on vision for catching prey and their eyes develop much more rapidly (fig. 5). Second-instar *Aeshna* may have 250 ommatidia in each eye, whereas *Calopteryx* at the same stage has only about seven (Corbet, 1962). Climbers require abundant submerged vegetation and clear water, and are therefore very susceptible to pollution. They are opportunistic feeders and may take many types of small Crustacea and insects as well as oligochaete worms, snails, snail eggs and even tadpoles and small fish. In the laboratory zygopteran larvae can be fed on water fleas (*Daphnia*): Thompson & Pickup (1984) found that eleventh-instar *Ischnura elegans* consume 10–20 *Daphnia* a day at a steady rate until the last 2 days before the moult. In the wild the food of larvae can be determined after their capture by microscopic examination of faecal pellets. The larvae of *Anax imperator* select food in the following order of increasing preference: gastropod snails, caddis larvae, mayfly larvae, Zygoptera larvae, waterboatmen (*Corixa*), chironomid midge larvae. Although they normally like chironomid larvae best and seldom take snails, this set of preferences can be altered by individual experience (Blois & Cloarec, 1985). Kanou & Shimozawa (1983) found that blinded *Aeshna* larvae could still catch prey by tactile means.

Climbers usually wait in readiness for their prey and then turn towards it slowly until it is in line with the head and at the right distance from the labial mask for a strike. Hungry larvae, however, respond to moving prey 20 cm away by stalking. If they are made to stalk an artificial prey which is then suddenly withdrawn, they perform a stereotyped behaviour pattern, first staring, then backtracking and finally rotating so as to face in a new direction. This fascinating behaviour, which looks like a dance of frustration, seems to ensure that they do not continue to pursue a prey they are unlikely to catch. Etienne (1978)

ommatidium: a unit of the compound eye

Fig. 5. Eyes of dragonfly larvae. (*a*) *Anax*, a visual hunter with large eyes and small antennae. (*b*) *Orthetrum*, a tactile hunter with small eyes and larger antennae.

(a)

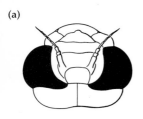

(b)

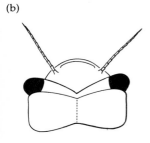

showed that the amount of back-tracking was directly
related to the effort expended in the original pursuit; she
did this by attaching a float to a larva which impeded its
forward progress and interfered with its pursuit of a prey.

Territorial behaviour in larvae

The larvae of some Zygoptera are known to hold and
defend territories against other larvae of the same species,
and they may thereby acquire better feeding sites (Corbet,
1980). The behaviour can be studied in the laboratory by
placing marked larvae (see chapter 6) in a tank with a
frame to climb on (thick netting or twigs). When an
intruder approaches there is first a prolonged and
motionless staring match during which the contestants
may curl their abdomens round so as to present the
leaf-like lamellae at the tip to each other. A confrontation
may end in some labial snapping before the intruder
retires (Rowe, 1980) (fig. 6). Harvey & Corbet (1985)
discovered that successful territory holders in *Pyrrhosoma
nymphula* were larger than other larvae and in turn they
gave rise to larger adults. Such adults, if male, may also
be more successful in holding territories and thus in
acquiring mates (see chapter 3), while if female, they
probably are able to lay more eggs. Thus successful larvae

Fig. 6. Territorial conflict in
zygopteran larvae. The
territory holder is to the left
and an intruder to the right.
(*a*)–(*e*) Successive stages in the
conflict. (After Rowe, 1980.)

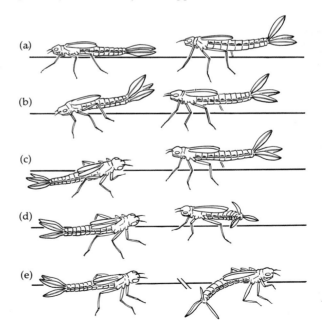

tend to give rise to successful adults. This work raises many intriguing questions about the nature of the conflict between rivals, about how it is conducted, and about what determines 'good' territories. Little is known about the existence of territorial behaviour in larval Anisoptera, but this would make an excellent topic for study needing no more than marked larvae, aquaria with supports and patient observers.

Larval respiration

The respiratory system of larvae consists of tracheae which are filled with air. The normal openings of the tracheal system to the outside at the spiracles remain sealed until the last larval instar. Larvae respire by means of gills which are filled with a mesh of tracheae lying just below a very thin cuticle (0.2–1.0 micrometres thick in Anisoptera), and which present a large surface area for the exchange of gases between the tracheae and the surrounding water (Komnick, 1982). The three lamellae at the tip of the abdomen of Zygoptera function as gills in this way although gas exchange can also take place across other regions of the body: hence after the removal of the lamellae, larvae can survive in well-aerated water by using the abdominal wall and, in later instars, the wing sheaths, as exchange surfaces. There may also be some exchange across the wall of the rectum, as in Anisoptera. When oxygen is short, intact larvae wave the lamellae and they may also climb to the water surface, or even into the air.

The 60–80 small gills of Anisoptera are housed in the rectal chamber, a part of the hindgut into which water is regularly pumped by abdominal muscles (fig. 7). The chamber also contains important glands concerned with the uptake of ions from the water. In *Aeshna*, expansion of the abdomen sucks water into the chamber under a negative pressure of about 0.5 cm water; the subsequent compression expels it forcibly under a positive pressure of 2–4 cm water through partly closed anal valves. In this way the re-uptake of the same water is prevented (Mill, 1974). By expelling water more violently, larvae can propel themselves along very rapidly. During such jet propulsion the abdominal pressure rises to about 40 cm water and the legs are folded along the body. Jet propulsion provides larvae with an effective means of escaping from predators, and it also may be used in making long migrations, for example to the water margin before emergence.

Rectal ventilation is practised by all anisopteran larvae. In some gomphids the last few abdominal segments are

Fig. 7. The rectal chamber, containing the gills, of an aeshnid larva.

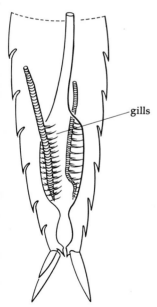

gills

greatly extended allowing the larva to burrow without sucking mud into the rectum. Buried *Cordulegaster* turn the end of the abdomen upwards for the same reason.

Rectal ventilation can readily be examined in the laboratory. If a larva is restrained in Plasticine, ventral side up, a small hook (made from an entomological pin) can be passed through a ventral plate of the abdomen and then joined with thread to a recording apparatus. After placing the larva in a shallow dish of water, ventilation can then be recorded using either an electronic transducer and recorder, or a lever and kymograph. In this way the amplitude and frequency of pumping movements can be compared under various conditions such as changed temperature. Alternatively a small water-filled tube can be introduced into the rectum and joined to a pressure transducer, allowing the pressures generated during pumping or with jet propulsion to be measured (Mill, 1974). Even without recording apparatus it is possible to learn a lot about ventilation by counting the number of pumping movements made each minute and seeing how this varies with temperature or activity. If a few drops of a dye such as methylene blue (or ink will do) are added to the water near an anisopteran larva, the movement of water in and out of the rectum can be watched. By persuading a larva to jet-propel itself through an ink cloud, a vivid impression of the force exerted can be obtained.

3 The adult

Emergence

Towards the end of larval life the compound eyes develop rapidly, the wing buds expand and the flight muscles increase in size. The respiratory rate rises and, a day or two before emergence, feeding ceases and the tissues are withdrawn from the terminal parts of the labial mask. At the same time many species select plant stems or other suitable emergence sites and sit with the head and part of the thorax out of the water, as though contemplating the aerial world they are about to enter, but in reality exchanging gases through the first spiracle, which can now be opened. Shortly before emergence the gills cease to function and rectal ventilation comes to an end.

In temperate climates libellulids and gomphids usually leave the water and emerge early in the day. They are then ready to fly when the sun warms them. Aeshnids commonly emerge during the night, but in cold weather they can postpone emergence until the following day. Zygoptera normally emerge during the day as well. Most species require a vertical support for emergence and they choose plants, rocks, trees or posts, sometimes travelling several metres from the water to find a suitable site. They have even been known to choose the trouser leg of an observing naturalist. In contrast gomphids can emerge and inflate their wings in a horizontal position, and they sometimes do so very close to the water margin, or on water lily leaves.

Emerging dragonflies are very vulnerable to predation from frogs, spiders, other dragonflies and birds and they may also be adversely affected by wind, rain and low temperatures. Corbet (1957) found a mortality of 16% in a population of emerging *Anax imperator* on a cold night. Small dragonflies may be able to fly within 30 minutes of emergence, but larger ones wait at least 1–2 hours and they usually postpone flight for longer. Natural selection has probably favoured dragonflies that can cast their skins rapidly, harden their new cuticle quickly, and remain hidden as far as possible while they do so.

Emerging dragonflies make fascinating subjects for study. Many photographs of emergence have been published, but in spite of this there are few accurate and detailed descriptions of the complex series of movements involved. Photographs do not depict movements, and photographers sometimes look as much at their equipment as at their subjects. There is therefore still a need for accurate and detailed accounts of emergence, particularly when they can be supplemented by the use of a video recorder and close-up lens. By using a mirror it is possible to record simultaneous views from two directions 90° apart.

Some temperate species overwinter in the last larval stage and emerge early in the season in well-synchronised bursts. These are known as 'spring' species, and examples are provided by *Pyrrhosoma nymphula*, *Gomphus vulgatissimus* and *Anax imperator*, which emerge in May or June. Many hundreds of *A. imperator* may emerge from a pond on a single night with resultant competition for emergence sites (Corbet, 1957). Likewise *Gomphus* may emerge in large numbers from a river during a few days in late May. Synchronised emergence ensures that many individuals are sexually mature together and it may allow the dragonflies greatly to outnumber their predators. However, sudden prolonged bad weather may decimate a population. In contrast 'summer' species overwinter in an earlier larval stage, or as an egg, and then emerge over a longer period later in the summer. *Aeshna grandis*, for example, emerges from June onwards, whereas *A. mixta* may sometimes not start to emerge until August and is on the wing until late October. Other summer species include species of *Sympetrum* and *Lestes sponsa*.

Exuviae

The presence of adult dragonflies at an aquatic habitat cannot alone furnish evidence that they breed there. Some dragonflies are great wanderers and they make many visits to unsuitable habitats. Finding cast skins (exuviae), however, confirms the completion of larval development. Exuviae can be collected among such plants as *Sparganium* (Bur reed), *Scirpus* (Club rush), *Iris* (Yellow flag) and *Typha* (Bulrush), but they must be searched for carefully. They will remain in position for some weeks in dry weather but may quickly disappear after rain or wind.

The species can usually be identified from exuviae (chapter 4), and the sex can be determined by looking for signs of the developing secondary genitalia of males. Softening the exuviae in water or 70% alcohol may make identification easier. Collections of exuviae throughout the emergence season can give information about the sex ratio at emergence as well as the species present at a habitat and their relative abundance. Sex ratios among mature adults at breeding sites are often strongly biased towards males because females spend much time elsewhere. Thus collecting exuviae is not only enjoyable and interesting, but it also gives information of scientific value.

Flight

Above all else it is the amazing powers of flight of dragon-flies which fascinate those who watch them, and since almost all their activities can be carried out on the wing, it is appropriate to give some thought to their flight mechanisms. Although the dragonfly flight system is primitive in the arrangement of flight muscles, the organ-isation of the cuticular plates of the thorax and the type of wings, the flight performance is unequalled in any other insect of comparable size (hawk moths may be close rivals).

A system of longitudinal veins arranged to give a fluted, corrugated structure provides the wing with stiffness in the long axis. A nodus or break half-way along the leading edge of each wing (fig. 8) allows greater twisting of the wing about the long axis and seems to be an important specialisation of dragonfly wings: it may also act as a shock absorber (Norberg, 1975). Numerous small veins spread across the wings forming a net which

Fig. 8. The fore and hind wings of dragonflies. (*a*) Anisoptera (*Orthetrum*), (*b*) Zygoptera (*Ischnura*).

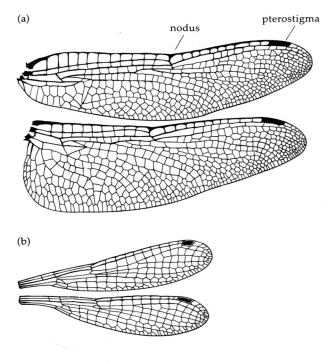

(a)

nodus

pterostigma

(b)

combines strength with flexibility, and dragonfly wings are no heavier per unit area than the wings of flies with very many fewer veins. May (1981) has shown that the wings of male Anisoptera are relatively longer and narrower in large species than in small ones, with consequent improvements in the generation of lift and the reduction of drag. Dragonflies are very economical in terms of the energy consumed per unit of distance travelled (the cost of transport), largely as a result of their efficient wing design. Some species have the posterior region of the hind wings broadened, which assists in manoeuvrability and in gliding. Those species which perch a lot have a relatively greater wing mass than those which habitually patrol in flight, and ground perchers (e.g. gomphids and *Orthetrum*), which may be susceptible to damage at take-off, have an even higher wing mass.

Most dragonflies bear a prominent spot, the pterostigma, near the wing tip, although it is reduced or absent in some Calopterygidae and Pseudostigmatidae (fig. 9). The pterostigma is thought to serve an aerodynamic function by adding weight to the tip, which may assist in twisting the wing at the top and bottom of the stroke (Norberg, 1972). In some species the pterostigma is brightly coloured and it may contrast with the wing colour (e.g. it is white in some black-winged species). It may therefore serve an additional visual function, perhaps in sexual interactions or as a wing-tip indicator.

In flight the fore and hind wings normally beat out of phase, but they show much independence of control and this accounts for the high degree of manoeuvrability of many species which can fly upwards, sideways and backwards as well as forwards. During banking and at

Fig. 9. Male *Calopteryx splendens*, perched with wings folded.

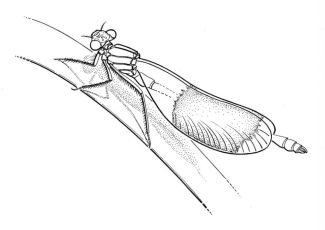

take-off *Aeshna* beats its wings in phase, developing more power (Alexander, 1984). Flight speeds are difficult to estimate in the field, but large aeshnids may patrol at speeds as low as 1–2 metres per second, and then accelerate after prey at perhaps more than 15 metres per second (54 kilometres per hour). The long abdomen of dragonflies gives some inherent stability, particularly in those gomphids and petalurids which have expanded lobes resembling fins on the posterior abdominal segments. Flight stability is maintained, however, mainly by behavioural adjustments to the visual field mediated by the eyes.

Wing stroke frequencies are low in Zygoptera, particularly in *Calopteryx* species: in *C. splendens*, for example, the frequency is only about 16 beats per second (Rudolph, 1976). In Anisoptera, however, it reaches 30–40 beats a second – higher than expected in insects of their size and weight (May, 1981). For example in free flight *Anax* beats its wings up to 31 times a second, while *Pantala* (a libellulid) does so at 39 times a second when fully warmed up, and such high frequencies contribute to the excellent manoeuvrability of dragonflies.

During a wing beat, the leading edge of the wing turns downwards at the top of the stroke and upwards at the bottom. This twisting of the wing in its long axis is an important part of the way by which lift is generated. Male *Calopteryx* use their coloured wings both for signalling and for flight. This has recently been filmed in *C. splendens* by Rüppell (1985) at 500 frames a second. In normal flight the fore and hind wings beat together; in territorial flight and in contests they also beat in phase but they hesitate at the top of the stroke (i.e. when the wings are back along the abdomen) which makes the black band conspicuous. In courtship flight the wings beat alternately at four times the normal frequency and with a reduced amplitude.

Another factor which makes dragonflies such excellent aerial acrobats is their unusually low wing-loading (mass per unit of wing area). In species with wings about 50 millimetres long, the loading is 0.037 grams per square centimetre, whereas in those with wings about 20 millimetres long it is only 0.018 grams per square centimetre. This may also enable them to compensate effectively for severe damage to one wing, or to fly using only one pair of wings as some species of *Calopteryx* do in courtship.

Most insects fold their wings backwards using a separate wing hinge mechanism. This is lacking in dragonflies, but because the wing bases are far back on the sloping thorax, Zygoptera can effectively fold their wings along the abdomen by elevating them maximally (fig. 9). This allows them to perch and walk among thick vegetation and even to descend below the water surface during oviposition. Anisoptera, in contrast, do not fold their wings in this way and they are very reluctant to walk, perching usually on exposed sites.

Hovering

Dragonflies are often seen to hover while active. They do so with the body horizontal and the wings moving diagonally with fast, small-amplitude strokes. *Aeshna juncea*, for example, hovers by beating its wings 36 times a second, with a stroke plane at 60° to the horizontal and with increased wing twisting at the top and bottom of the strokes which is probably of major importance in generating lift (Norberg, 1975).

Hovering is probably more expensive than forward flight in terms of fuel used per unit time, and it may also bring thermoregulatory problems (see below). Nevertheless males often hover in their territories, or when they search for females elsewhere, and so do some females when they inspect oviposition sites. Hovering may not only serve to advertise the presence of a male in his territory, but may also allow an airborne insect to see better and to detect movements more readily than if it were in forward flight, particularly at low light intensities. This may be one reason why males spend a considerable amount of time hovering in their territories.

Flight metabolism

Flying dragonflies increase their metabolic rate many times compared with the resting rate, and they have a high power output particularly during hovering. There is thus a high demand for fuel and oxygen in flight. Most flight is probably fuelled by carbohydrate stores, but long migrations may depend more on stored fats, as in other animals. Oxygen is supplied to the flight muscles through a highly specialised tracheal system which allows the flight muscles to be ventilated by the thoracic flight movements through open spiracles (autoventilation) (Weis-Fogh, 1967). When the wings are in the mid-position (i.e. horizontal) the right and left halves of the thorax are forced slightly apart, a movement permitted by the flexible fold in the dorsal midline of the thorax. By moving the wings up and down passively by hand you can see how this fold opens and closes. The two halves of the thorax come together at the top and bottom of the stroke, and the thorax is therefore ventilated twice per wing-beat cycle. Weis-Fogh has calculated that in an

Aeshna juncea weighing 1 gram, autoventilation can contribute about 8.35 microlitres of air per wing stroke, which is very much more than can be ventilated by the abdomen.

In many insects the tracheal system ends in exceedingly fine tubes, the tracheoles, which push into the fibres of the flight muscles and come to lie amongst the mitochondria. Thus oxygen can be brought very close to the powerhouses of the flight motor. However, in some dragonflies the tracheoles do not seem to indent muscle fibres in this way but terminate on their surface. Probably for this reason the muscle fibres of dragonflies are slender, being not more than 20 micrometres in diameter, and the large mitochondria extend to the edge of the fibre. Weis-Fogh argued that the design of the dragonfly flight system is pressed to its limits and that it could not provide adequately for an insect much larger than modern aeshnids. Perhaps, therefore, the giant meganeurids of the Carboniferous flew less actively, gliding for much of the time, or perhaps they possessed adaptations not found in present-day species.

Thermoregulation

Adult dragonflies are very dependent on warmth and sunshine for their activity, and on cold grey summer days you should not expect to see much behaviour in the field. Moreover in very hot conditions their rates of activity can lead to overheating. It is not surprising, therefore, to find elaborate behavioural, morphological and physiological adaptations that help them to regulate body temperature. Flight generates metabolic heat and in larger species this warms the thorax faster than convection can cool it so that the temperature rises; the flight muscles are found to function best at a high temperature. Large aeshnids are unable to fly with a thoracic temperature below 30–40 °C. Some small libellulids (weight 100–200 milligrams) can do so at about 16 °C, although many prefer not to fly until the thorax reaches about 24 °C: at the higher temperature flight is faster and more efficient (Vogt & Heinrich, 1983). In contrast most Zygoptera can fly with a much lower thoracic temperature.

Many large species must therefore warm the thorax to a temperature well above ambient before they can fly, and they do this by behavioural and physiological means. Behavioural warming depends on basking in sunlight so as to intercept the maximum radiation (fig. 10a). They perch at right-angles to the sun's rays in sheltered positions, sometimes gaining additional heat by sitting on hot rocks or the earth. Black, brown or red species absorb

Fig. 10. (*a*) *Sympetrum* basking
in the morning sun and
casting a large shadow. (*b*)
Sympetrum perching at noon
on a hot day and casting a
minimal shadow.

(a)

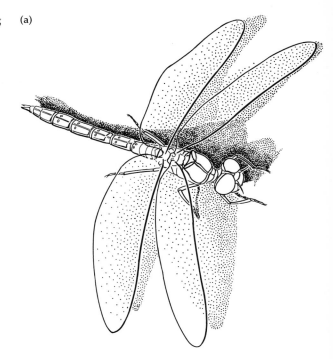

(b)

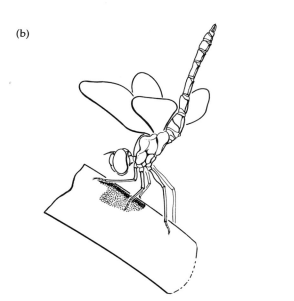

more radiation than others, and regions of transparent
cuticle on the dorsal side of the thorax or abdomen may
allow them to heat up as does a greenhouse. Adaptations
of this nature are beautifully exemplified by northern
species such as *Libellula quadrimaculata* with its brown and
black coloration, transparent areas of cuticle and broad
abdomen.

Some blue zygopterans and aeshnids may be able to
change colour according to the temperature. Their blue
colour depends on light scattered by tiny particles (about
0.23 micrometres in *Enallagma*) in the epithelial cells.
Below about 10 °C the particles are moved deeper within
the cells, perhaps as a result of the activity of a hormone,
and the dragonfly turns grey. The change can readily be
seen by placing *Enallagma cyathigerum* in a fridge for 30
minutes. It is thought that the grey colour allows them to
heat up faster in the sun. Surprisingly, the similar blue of
Coenagrion puella does not change on cooling to 5–10 °C.

Some species of *Libellula* and *Orthetrum* appear blue
due to a wax bloom on the outside of the cuticle. This
reflects radiation and tends to be more extensive in
species of warmer climates. Thus, whereas only the
abdomen of *Orthetrum cancellatum* and *O. coerulescens* is
blue, both abdomen and thorax are blue in the south
European species *O. brunneum*. Overheating in perched
species is avoided by behavioural means: libellulids may
perch with their abdomen pointed towards the sun at
midday (fig. 10*b*) casting a minimal shadow (thus warming up less than when perched at right-angles to the
radiation), or they perch higher on plants or in the shade.
Some species are adept at choosing sun-dappled regions
so that they gain just the right amount of radiation (May,
1978).

Physiological means of gaining heat depend on rapidly
shivering the flight muscles while perched. Shivering or
wing-whirring is commonly observed in aeshnids and
other large dragonflies before take-off and in females
while they oviposit among plants. The wings initially
make small vibrations caused by the simultaneous contractions of the flight muscles (Pond, 1973). By this means
Aeshna can rapidly warm up the thorax to 30 °C, starting
from an ambient temperature of 12 °C, at a rate of 3–7
degrees per minute (Vogt & Heinrich, 1983). Heat reten-
tion is improved by a layer of insulating airsacs which
almost surrounds the flight muscles of most Anisoptera,
and in a few species (e.g. *Brachytron pratense*) also by a
coat of hairs on the outside. Only a few species of libellulids use this method of heat gain, and they do so at
times when there is no sun to bask in. An example is the
small libellulid *Sympetrum depressiusculum*. In southern

France the males of this species awake at dawn, wing-whirr to make flight possible and then go in search of females. Having formed a tandem with a female they then settle for 2 or 3 hours until the sun makes it warm enough for mating and oviposition. It is a remarkable experience to hear thousands of males all wing-whirring at dawn in rice fields in the Camargue where this species sometimes abounds. Similarly in early autumn *Anax junius* in North America wing-whirr at dawn, then fly around the bushes they have roosted on and sit basking in the rising sun (Corbet, 1984).

Large dragonflies which patrol during the middle of the day may tend to overheat. They can avoid overheating, however, by making progressively longer glides between bursts of wing beats as the temperature rises – a style characteristic of many aeshnids. They may also increase the rate of blood circulation through the thorax, the blood carrying heat to the abdomen from where it is dissipated. Heinrich & Casey (1978) artificially heated the thorax of aeshnids and found that the transfer of heat to the abdomen could be blocked by ligaturing the heart, thereby providing good evidence for the thermoregulatory function of the blood. No heat transfer mechanism of this type has, however, been found in libellulids.

You can examine flight in a tethered dragonfly in the laboratory in front of a wind tunnel, or a fan. Glue or wax the underside of the thorax, behind the legs, to a small stand made from a matchstick held in a clamp. (Copydex can be used for small Zygoptera. A mixture of beeswax and resin should be used for large dragonflies.) Short bursts of flight can usually be initiated by gently pinching the abdomen. The wing movements and the beat frequency can be examined with a stroboscope or with flash photography. With a stroboscope you can also see how the wings are twisted at the top and bottom of each stroke.

Recording thoracic temperatures in the field is difficult. The most successful method has been to catch adults with a net and immediately to insert a minute thermocouple into the thorax. This must be done within a few seconds of capture before appreciable cooling has occurred, but it leads to the death of the dragonfly. However, much thermoregulatory activity can be observed without direct measurements of thoracic temperature; a knowledge of environmental temperatures often helps to explain dragonfly behaviour.

Here are some problems concerning thermoregulation that you could try to solve:

1. How is the duration of wing-whirring before flight in large aeshnids related to the ambient temperature?

2. On a hot day in late summer how do the perching positions of *Sympetrum striolatum* vary during the day? What is the angle between the long axis of the body and the sun's rays at different times?
3. To what extent does a dragonfly's colour affect its temperature in the sun? A preliminary answer might be obtained by making a paper model of a dragonfly coloured to resemble the male of a particular species. Place it where you have seen the dragonflies perching and measure the temperature inside the model. Compare the temperatures of white and black models similarly placed.
4. At what temperatures are Zygoptera unable to fly when disturbed? How do they and other dragonflies respond to disturbances when too cold to fly?
5. Does the hypothesis that dragonflies glide more when they are hotter apply to all species? To answer this the following observations are suggested. Time the duration of glides and the proportion of time spent gliding in hawking or patrolling aeshnids. Note the time of day, the air temperature and the amount of cloud cover. Compare the flight performance under different conditions.
6. If a male *Enallagma cyathigerum* is placed in a fridge at 5–10 °C for an hour, and then taken out and photographed alongside another male kept at about 20 °C, one can compare the colours, and see how long it takes them to become identical as the first male warms up. Do other blue species also show colour changes when cooled?

Some useful hints on how to make measurements of important environmental variables can be found in a little book by Unwin (1980).

Vision

Dragonflies have very large eyes relative to body size, and more than 80% of their brain is devoted to the analysis of visual information. Because vision dominates their behaviour, studies of their other senses have been neglected, but touch receptors and chemoreceptors probably also have important roles to play. The possible involvement of chemical messages (pheromones) too, at least in close-up interactions, cannot be discounted: the small antennae of dragonflies bear sensory structures resembling those which detect chemicals in other insects.

The head of an anisopteran consists largely of eye. Not only are there two enormous compound eyes, but three simple eyes, the ocelli, can also be seen on the upper part (fig. 11). The forward-looking ocellus is partly hidden

Fig. 11. The eyes of dragonflies. (*a*) Head of a zygopteran: note the three prominent ocelli. (*b*) Head of a libellulid to show the large compound eyes and the ocelli. The central ocellus is concealed below a bulge of cuticle. (*c*) A libellulid head from the side to show the transition between large dorsal ommatidia and small ventral ones in the compound eye.

(a)

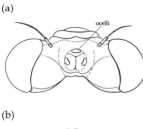

(b)

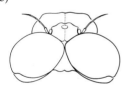

(c)

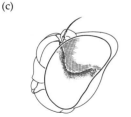

under a cuticular visor which ensures that the field is centred on the horizon ahead. The lateral ocelli look sideways and upwards (Stavenga and others, 1979). Each ocellus has a large curved lens covering several hundred light-sensitive cells. The ocelli have a wide angle of acceptance, do not focus an image and are extremely sensitive to small changes of light intensity. They produce fast responses in the flight motor system, and are concerned with maintaining stability in flight (Simmons, 1982).

The compound eye is made up of ommatidia, each ommatidium being a visual unit consisting of a lens system and a group of eight photoreceptors. A large *Anax* may have 30 000 ommatidia in each eye. The photoreceptors are arranged in tiers and they send their axons to the optic ganglia of the brain. An ommatidium is to some extent a separate visual detector responding to the intensity of light in its part of the visual field. The brain assembles the inputs from thousands of ommatidia to form a mosaic image from which it extracts relevant information. Dragonfly eyes are extremely sensitive to movement and they point in almost all directions giving virtually a 360° visual field in the horizontal and vertical planes.

The ability of compound eyes to resolve fine detail (acuity) is less good than that of vertebrate eyes. In insects high acuity depends on a large ommatidial lens diameter (which reduces diffraction interference), a small angle between neighbouring ommatidia and a small angle of light acceptance by each ommatidium. The compound eye has regions specialised for high acuity, such acute zones being sometimes detectable as areas where the ommatidia are larger and the eye surface flattened, the latter reducing the inter-ommatidial angle. If you look at the eye of a living dragonfly (e.g. *Enallagma*), a black region, the pseudopupil, can be seen below the lenses where all light on the visual axis of the observer is absorbed. A large pseudopupil indicates the presence of a region of high acuity. In Zygoptera the pseudopupil is large in regions which look forwards and upwards, but small in those looking downwards (fig. 12). In libellulids there may be two or three acute zones, and there may also be a sharp transition between the small lower ommatidia and larger ones which look upwards (fig. 11*c*). Dragonflies often approach prey and rivals from below and behind, thereby using the region of the eye with the best acuity.

There is both behavioural and physiological evidence that dragonflies possess colour vision. In the eye of *Sympetrum*, Armett-Kibel & Meinertzhagen (1983) have described four types of photoreceptor which have maximum sensitivities for ultraviolet, blue, green and orange

Fig. 12. Pșeudopupils in the eyes of a zygopteran. Large pseudopupils indicate regions of the eye specialised to see fine detail. (*a*) The eye from below showing a small pseudopupil. (*b*) The eye from in front showing a large pseudopupil.

(a)

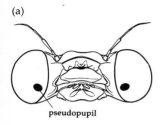

pseudopupil

(b)

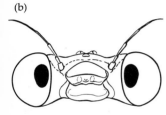

light respectively. Colour vision depends on analysing the inputs from such receptors in the brain. Many of the ultraviolet receptors are in the upper region of the eye, but some dragonflies have additional ones looking downwards which respond perhaps to ultraviolet reflected by water or by other dragonflies. Dragonflies also show sensitivity to polarised light. They use the dorsally located ultraviolet receptors for this purpose, as do bees and ants, and the information may be used in orientation. Some of the orange receptors in *Sympetrum* also show sensitivity to polarised light: they may be specially designed to enhance the contrast between water and other males, which tend to be red or orange in this genus.

Dragonfly eyes are sensitive to small movements, and they may mediate escape responses when a jerkily moving or expanding object appears in the visual field. A different type of response occurs when prey or another dragonfly is seen, and the compound eyes then control the complex tracking behaviour which may follow. Movements of the whole visual field sometimes produce compensatory (optomotor) responses by which an insect can remain stationary in relation to its environment. These are seen not only in adults, for example when hovering, but also in larvae in a water current.

Territorial males may return again and again to the same perch, and they may move between a roost and a territory on several successive days by the same route. They therefore seem to have a good visual memory of their locality. Moreover when a new feature is introduced into the familiar territory of a male, the dragonfly may come to inspect it. Robert (1958) has described the curiosity of a male *Aeshna cyanea* which, when he stood near its flight path, came to see what 'strange monument' was in its territory. In bees it is thought that a map of the surroundings may be memorised in the mushroom bodies – a region of the brain containing many tens of thousands of tiny neurones. However, dragonflies have very small mushroom bodies, and so perhaps they cannot learn so much.

Both larval and adult dragonflies are able to estimate their distance from prey. Stereoscopic vision may be used by adult Zygoptera and by many larvae for this purpose; it depends on using both the eyes and on their being some distance apart (fig. 11*a*). However, the eyes of most Anisoptera meet in the dorsal midline and such species may depend more on parallax and other means for judging distances. Knowledge of the likely size and flight speed of prey or of possible mates may help in this and enable them to make accurate interceptions in mid-air.

Dragonflies have not only high acuity but also good

sensitivity at low light intensities, and this allows some species to feed at dusk on small flies at times when the human observer can barely see the dragonflies, let alone their prey. The long ommatidia, large-diameter lenses and appropriate pigment movements may all enhance sensitivity. Long ommatidia also help to improve resolution.

Feeding behaviour

Dragonflies are opportunistic feeders, taking whatever suitable prey is abundant. They may congregate in considerable numbers when termites or ants are flying, or near swarms of mayflies, caddisflies or gnats. Sometimes they seem to anticipate the presence of prey, arriving in the right place at the right time, and learning may play a role in improving their hunting success. Maturing dragonflies feed intensively: adult females may also require much food when they are developing eggs, and males when they are active on their territories. Fried & May (1983) have suggested that sometimes food shortages may limit sexual activity. Males which hold territories for many hours routinely feed in residence, but those which are territorial for shorter periods usually do not feed there. Dragonflies do not hunt in cold weather and prolonged cold spells in summer may cause heavy mortality due to starvation.

Most small libellulids perch on prominent sites while on the look-out for prey and then dash out when a suitable insect flies past. In this way such perchers search in only a limited area and they must keep warm enough by basking so that instant flight is possible. Hence perchers choose sites in the sun and do not hunt by this means at the beginning and end of the day. Good examples in Britain are provided by *Orthetrum* and *Sympetrum* species. While perched, some libellulids keep the wings depressed and twisted in readiness for the upstroke and instant take-off (fig. 13).

Larger Anisoptera normally feed by patrolling along bushes, hedges, woods and elsewhere. This method, typical of Aeshnidae, allows them to hunt in a large area but it has a correspondingly greater energy cost. Some libellulids also turn to patrolling late in the day when it is too cold to hunt from a perch. They may then join aeshnids in an intensive period of activity towards dusk, taking advantage of the numerous insects which swarm at this time on warm summer evenings. Feeding at dawn is uncommon in temperate species, but it occurs in some tropical ones.

Zygoptera, being less limited by temperature, can

combine patrolling and perching techniques. *Ischnura elegans* is adept at picking up aphids and other small insects off plants, hovering first and then suddenly zooming in on a target on a leaf or stem. Sometimes a dark mark on a stem elicits an attack, indicating that prey movement is not essential for recognition. Tropical pseudostigmatids take spiders from their webs while aeshnids may gather up small frogs from the ground. Food is grasped by the forwardly directed legs which form a basket, and it is usually eaten on the wing by patrollers, but after settling by perchers.

Various methods have been used to examine the quality and quantity of food eaten. Accurate field observations can supply much useful information, particularly in perchers, when the prey can often be identified through field glasses after the dragonfly has settled. Analysis of gut contents can indicate more extensively the range of prey species, but it may be difficult to identify the small fragments of cuticle more closely than to insect family. By keeping starved adults under constant conditions in the laboratory and then feeding them by hand it is possible to measure the rate at which the gut handles food. Hand-feeding with flies or other insects is easy in mature dragonflies but it is more difficult to get immature individuals to accept food. Pieces of food should be offered by touching the palps; this usually excites chewing movements.

By careful observation and measurement, Higashi (1973) estimated the food intake of *Calopteryx* and *Sympetrum* in Japan. He found that *Calopteryx* made 150 feeding flights in a day, 43% of which were successful and provided 6 milligrams of food (11% of the dragonfly's body weight). *Sympetrum* made as many as 320 feeding flights

Fig. 13. A libellulid perched with wings lowered and twisted ready for instant take-off.

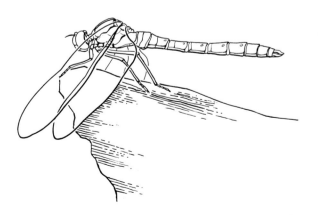

a day, 51% of which were successful and provided 12 milligrams of food or 14% of the dragonfly's weight. It would be particularly valuable to know how such intakes vary in dragonflies of different ages and sexes under a variety of conditions.

Maturation

For some time after emerging, most dragonflies avoid water and spend their time feeding and maturing. The cuticle thickens, the flight performance improves and the gonads develop. Aeshnids may migrate several kilometres from their emergence sites whereas many Zygoptera travel much less or not at all, and immature *Calopteryx* may remain gregariously together, close to water. Maturation can take from a few days to several weeks in Zygoptera depending on the species and on the temperature; in Anisoptera it is usually completed in about 2 weeks.

Towards the end of the maturation period males may take on their sexually distinctive colour patterns, while in most species females show less colour change although they may darken as they get older. In a few species old females become coloured like males, and this may possibly reduce the amount of attention they receive from males and allow them to oviposit undisturbed.

aestivate: to pass the summer in an inactive state

The development of male coloration is greatly delayed in a few tropical species which emerge at the beginning of the dry season and aestivate as adults until the next wet season. Similarly some temperate lestids retire to woodlands away from water for 2–3 months after emergence and become sexually mature only late in the summer. It may have been this adaptation which allowed another lestid genus, *Sympecma*, to evolve the ability to hibernate regularly. These small brown dragonflies, which are found in Holland and further south, hide away among leaves in the autumn and are not normally active again until the following spring, except during unusually mild winter spells.

hibernate: to overwinter in an inactive state

polymorphism: the presence of two or more forms of one species

Inherited colour polymorphism is common among the females of several zygopteran species such as *Enallagma cyathigerum* and *Ischnura elegans*. Males are not known to show a preference for a particular colour morph, and the function of the polymorphism is not understood.

The length of adult life varies considerably with temperature, food, population density and the abundance of predators. In *Calopteryx virgo* in Japan, maturation normally takes up to 11 days and the period of sexual maturity lasts on average 8–10 days in males and 11–13 days in females (Miyakawa, 1982). Fincke (1982) found the mean life expectancy of *Enallagma hageni* in America to be about 12 days and the maximum 20 days, while the mean for *E. cyathigerum* in Britain is the same but the maximum is 39 days (Parr, 1976). Maximum life expectancy for some adult Zygoptera, including the immature phase, is 7–9 weeks, while for species of *Aeshna* it is 8–10 weeks (Corbet, 1980). Some tropical species, however, can live for up to 9 months.

It is difficult to get reliable estimates of the age of dragonflies unless they can be marked at emergence. When this cannot be done, an indication of age in some species can be obtained from their progressive colour changes, although as already indicated these vary with temperature and food abundance. An alternative method is offered by the observation that cuticle continues to be laid down in a series of daily layers for up to 2–3 weeks after emergence in large dragonflies. Thus by counting the rings in a sectioned leg using a polarising microscope, a minimal estimate of the number of days since emergence can be made. However, caution should be exercised as this method still needs to be calibrated with dragonflies the age of which is independently known (Veron, 1973; Neville, 1983). Another indicator of age is the presence and size of water mites attached to adults, some of which may be parasitic as well as hitching a lift to the next aquatic habitat (Münchberg, 1982). The state of the gonads, the presence of puncture marks on the eyes of females made by the males' claspers during mating (see below), the wear and tear of wings and the presence of water marks on females which have oviposited can all be used as indicators of maturity but not of precise age (Corbet, 1962; Johnson, 1973).

Migrations

Migrations in which adult dragonflies, sometimes in very large numbers, travel many hundreds of kilometres are regular features of the life histories of a few tropical species, but migrations occur less commonly in temperate regions. In Africa and elsewhere, *Pantala flavescens* is a well-known seasonal migrant tending to arrive at the onset of wet weather in huge aggregations of many millions. *Hemianax ephippiger* is a less regular African migrant known also to breed in southern Europe. Both species are rare visitors to Britain. Migrating dragonflies sometimes travel at great heights (Corbet, 1984), and they have been seen at 6000 metres in Afghanistan (Wojtusiak, 1974). Longfield (1948) has described large numbers of *Sympetrum striolatum* in Britain appearing like 'drifting snow', and similarly large numbers of other species of trematodes: parasites related to liver flukes. The larval stages may occur in invertebrates

Sympetrum are not uncommon in parts of Europe. Many *Sympetrum* individuals can also be seen in the autumn flying south through passes in the Pyrenees in company with other insect migrants. *Libellula quadrimaculata* appears in very large aggregations about every 10 years in parts of northern and eastern Europe, and Dumont & Hinnekint (1973) have suggested that one cause of such periodic migrations is the build up of the trematode parasites of the larvae in their aquatic habitats.

If you are lucky enough to spot a migration of dragon-flies, it is important to record as much information as possible. Note the species present, remembering that several species may travel together; note also their approximate height above the ground; the direction, time and date of the flight; and the prevailing weather con-ditions, including wind direction and strength, amount of sunlight and the temperature. If some of the migrants are caught their identity can be confirmed and they can also be dissected so as to record the state of their gonads, the amount of stored fat and the presence of food in their guts. Estimating numbers in a swarm is very difficult, but you can try by counting the numbers that cross 20–30 metres of a hedge or wall each minute and by noting the total time for which the swarm is seen. Alternatively photographs may give an idea of the swarm's density.

Predator avoidance

Adult dragonflies are prone to attacks from a variety of terrestrial predators including birds, lizards, frogs, spiders and other large dragonflies: for example in southern France a migratory bird, the bee eater, takes many dragonflies and when feeding their young they may specialise on tandem pairs of *Sympetrum*. When they approach water for reproduction dragonflies may also be attacked by aquatic predators such as fish, water bugs and water spiders. *Dolomedes* (wolf spiders) run at high speed across the water surface and sometimes grasp ovipositing Zygoptera or even libellulids. Water bugs such as naucorids and notonectids may grab the abdo-mens of endophytic species (p. 47) as they probe below the water surface, and the water scorpion *Nepa* has been seen to seize a male *Calopteryx* when the latter was perched on plants at the water surface. Spiders' webs near water often contain many zygopteran wings, and libellulids are regularly caught in the webs of *Argiope* in southern Europe. Even harvestmen can seize immature zygopterans.

Dragonflies are well equipped, however, to avoid predation. They have very fast visual responses and agile

Fig. 14. The roosting position
of a coenagrionid along a grass
stem. (After Askew, 1982.)

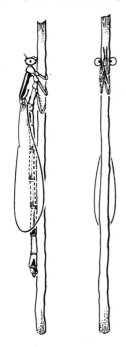

Fig. 14. The roosting position of a coenagrionid along a grass stem. (After Askew, 1982.)

flight, and when active can usually avoid all but the swiftest predators. In flight some species follow a lilting or bobbing path which may make their capture more difficult, while zygopterans commonly fly extremely close to the water surface. Pairs in tandem or in copulation are attractive targets as already indicated, and so at least in some species these activities are carried out rapidly. Some Zygoptera choose to roost on stems just thick enough to hide behind with the eyes protruding (Askew, 1982), and if it is too cold to fly they move round to the other side of the stem when danger threatens (fig. 14). Likewise libellulids sometimes roost in dense stands of rushes (*Juncus*) and drop down among the spiky leaves when alarmed. When caught, some aeshnids bend their abdomen round apparently with aggressive intent, and *Onychogomphus forcipatus* can pinch with its enormous claspers, while most dragonflies attempt to bite their attackers.

Although males are often conspicuous, immatures and females are sometimes more cryptically coloured and may also have disruptive patterns which make them difficult to detect when perched. Mimicry is not a means of protection employed by many dragonflies, but the brown and yellow pattern on the abdomen of female *Libellula depressa* resembles a hornet's markings and may deceive a predator.

Reproduction

It is the reproductive behaviour of dragonflies which above all attracts the attention of biologists, not only because it contains unusual features such as the employment of secondary genitalia, but also because it exemplifies important and widespread biological principles such as territorial behaviour and sexual selection. We will deal first with the sexual structures of dragonflies, and then with their sexual behaviour.

Males differ from females in the following ways:

1. The male thorax and abdomen are often more brightly coloured than those of the female, blues and reds being the commonest colours. In some genera, such as *Calopteryx*, the wings are also strikingly patterned, while the males of African Chlorocyphidae possess brightly coloured legs. Such colours may be used in the competition between males for territories and mates. In other species in which the sexes are similarly coloured, territorial behaviour may be absent (e.g. *Cordulegaster boltonii* and some gomphids) or less sustained (e.g. *Cordulia aenea*).

Fig. 15. Male dragonflies holding females in tandem. (*a*) A zygopteran male holding a female by the front part of the thorax. (*b*) A male aeshnid holding a female by the head.

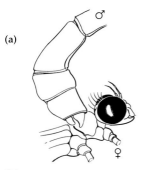

(a)

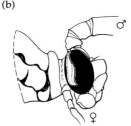

(b)

2. At the tip of the abdomen there are three or four structures, the claspers, with which males grasp females in tandem. The male's claspers grip the front of the female's thorax in Zygoptera, and the female's head in Anisoptera (fig. 15), before and during mating, and sometimes also during oviposition.

3. The base of the hind wing of male gomphids, peta-lurids and some aeshnids is sharply angled and has a strong marginal vein. This is associated with the presence on the second abdominal segment of aur-icles; these are small, often toothed, structures of unknown function (fig. 16). It has been suggested that the auricles make contact with the ovipositor during copulation, or that they support the male's hind legs in manipulating the female genitalia, or even that they act in conjunction with the angled hind wings in flight control or in sound production, but none of these proposals is supported by evidence. Accurate obser-vations on males at the start of copulation might establish their function but mating begins in the air and the details are hard to see. A close examination of the external and internal structure of the auricles is also called for to see whether, for example, they are equipped with sensory endings.

4. The primary genitalia of the male lie within the pos-terior segments of the abdomen as in other insects. The secondary genitalia are unique to the Odonata and are formed by the ventral plates of the second and third abdominal segments (figs. 17 and 18). They are employed as an intromittent organ during copulation, and are thus comparable in function to the pedipalps with which male spiders introduce sperm into females. Much speculation has been given to their evolution. Carle's (1982) view is that in an ancestor, a spermatophore (a sac containing sperm) was placed on the ground and then picked up by the female, as occurs today in some primitive arthropods. At the same time the male clasped the female on the head or

Fig. 16. (*a*) Ventral view of the large auricles (arrow) and angled hind wing (arrow) of a male *Cordulegaster*. (*b*) The female with no auricle and a rounded hind wing.

(a)

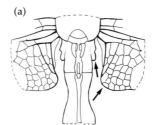

(b)

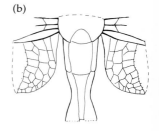

Fig. 17. The secondary
genitalia of a male zygopteran.

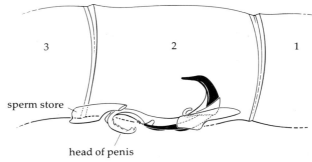

Fig. 18. Penis of a libellulid.
(*a*) Uninflated, (*b*) inflated.

(a)

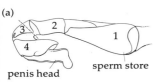

(b)

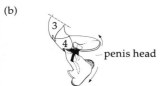

homologous structures:
structures which share a
common evolutionary origin

thorax to avoid being eaten by her; tandem pairing is
thus considered to have preceded the formation of the
secondary genitalia. Later the male placed the spermatophore on his own abdomen from where the
female picked it up, and he then evolved structures to
hold and finally to introduce it into the female. However, although spermatophores are known in several
insect orders they occur in no modern dragonfly and
they must therefore have been lost at an early stage
according to this theory. Modern aeshnids do transfer
sperm in bundles but these are not equivalent to
spermatophores.

The structure of the secondary genitalia

In Zygoptera the penis consists of a long, hard structure
with a flexible head developed from the second abdominal segment (fig. 17). The third abdominal segment forms
a sperm store (seminal vesicle) into which sperm is trans-
located from the primary genitalia shortly before copulation when the pair is in tandem. It is not known how
long sperm can survive in the store or whether they
receive nutrients there. This could readily be investigated
by catching a male zygopteran about to copulate and then
squeezing out small amounts of sperm at intervals of
several hours from the store onto a slide. Living sperm
can easily be seen swimming in a drop of 1% saline under
the microscope with a ×40 objective.

In Anisoptera the penis has evolved as a three-seg-
mented extension of the sperm store on the third segment, while a structure homologous with the zygopteran
penis (the ligula) serves to lever the penis posteriorly into
the female (fig. 18) (Pfau, 1971). The sperm store also
contains a separate reservoir of liquid, and the liquid can
be driven into the penis by muscular contractions: this

causes bristly sacs and other structures on the last segment to be expanded (see p. 43). The genitalia also include one or two pairs of hamules, claw-like structures which can grip hold of the female during copulation.

There is a great deal of variation in the copulatory structures in different species and more so between genera and families. This aspect, long made use of by taxonomists, only now begins to be understood functionally and in terms of sexual selection, as is discussed below.

The structure of the female reproductive system

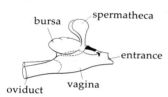

Fig. 19. The reproductive organs of a female libellulid.

The reproductive organs extend throughout most of the female abdomen, which in many species is thicker than that of the male. The ovaries mature successive batches of eggs over a long period. The paired oviducts lead into a muscular vagina from which sperm storage organs are given off on the dorsal side (fig. 19). The storage organs comprise the bursa and the spermatheca: the latter is single in many Zygoptera but paired in Anisoptera. The spermathecae may possibly act as long-term stores for sperm while the bursa is an immediate store, but there is much variability between species and no firm evidence about the separate functions of the two structures. That females can store viable sperm for 2–3 weeks has been shown by keeping them in captivity after mating, when they continue to lay fertile eggs. A female may receive from a single mating sufficient sperm to fertilise most of the eggs she will ever lay, although in the wild she may copulate with many males. We shall consider the possible reasons for this later.

Female Zygoptera and aeshnids bear prominent ovipositors. The ovipositor is formed from a pair of appendages on the eighth segment and two pairs on the ninth. With it a female can saw into plant tissue and then insert her eggs. In Libellulidae and some other Anisoptera, the ovipositor is greatly reduced, consisting of a small plate and some sensory structures, and eggs are not placed within plant tissue.

Having looked briefly at the reproductive equipment of the two sexes, we can now consider their reproductive behaviour.

Territorial behaviour

The relatively small size of many dragonfly territories around a pond or along a stream makes them excellent subjects for studies of territoriality.

A territory is a defended area where an animal shows

site attachment and is dominant, driving out intruders
(fig. 20). Among dragonflies there is a spectrum of types
of territoriality, ranging from those species in which
males are strongly and consistently territorial for long
periods – some males of *Calopteryx* and *Orthetrum* may
occupy the same territory for up to 3 weeks while
examples of tenure for 3 months are known in tropical
Zygoptera – to those in which territories are held for
shorter periods, sometimes only for a few minutes. Many
species are altogether non-territorial. The behaviour
within a species may vary considerably according to the
habitat, density of males, the time of year, the climate and
the age of individuals. Thus what is reported for a species
in one region need not always apply elsewhere. The
behavioural variability shown by dragonflies is fascinating and the observer must always be ready for the unex-
pected and exceptional.

   Orthetrum coerulescens stakes out territories along
stream banks about every 10 metres (Parr, 1983), whereas
O. cancellatum sometimes holds territories which are up to
50 metres long (Krüner, 1977); in contrast the territories of
some small dragonflies may be less than 1 metre in extent.
In some species the territory can be subdivided into a
central zone where the male spends most of the time and
wins all fights, and an outer region where he is less

Fig. 20. The territory (dashed
line) of a male *Calopteryx*.
(After Heymer, 1973.)

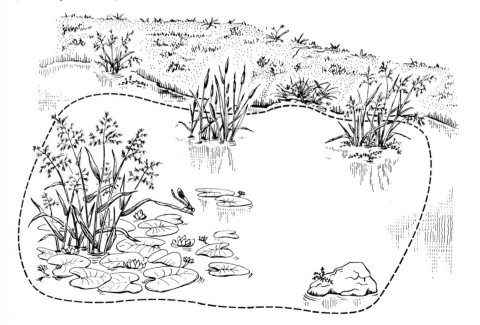

aggressive but which he patrols periodically. A territory on a stream may be confined to a linear strip of bank, but at ponds and rivers it may comprise a broader area. In either case it will include oviposition sites attractive to females.

Males resident in their territories usually divide their time between perching and flying. *Orthetrum coerulescens* flies for about 19% of the time, most flights being patrols up and down the territory but some being approaches to intruders. In contrast some aeshnids may fly continuously in their territories, hovering frequently, and they may follow the same flight path many times, settling only outside the territory.

As the density of males increases, often towards midday, territories in some species tend to become restricted in size and the amount of interaction between males may decline (Pajunen, 1962). Aggressive interactions, often containing ritualised movements in which males display brightly coloured abdomens (see below), may escalate into fast chases with clashes which sometimes knock one of the males into the water. In some species large males are more successful than small ones in holding territories, but in the American *Calopteryx maculata*, residents win almost all contests although they are not necessarily the largest individuals (Waage, 1983). Intruders seem only to check that a territory is occupied, and when occasionally a fight persists for much longer than normal, Waage suggests that each male believes he is the resident.

In some species there is a rapid turnover of males at the water, each holding a territory for only a short time. The most successful males will hold territories when females are most likely to arrive (Campanella & Wolf, 1974). Alternatively a male may occupy several territories in turn, remaining in each for a relatively short period. For example *Anax imperator* remains in residence in one territory for 93 minutes at the most, but usually for less time (Consiglio, 1976), while *Aeshna cyanea* holds a territory for about 40 minutes during which its aggressiveness and the amount of hovering decline; it then may move to another territory and return later in the day to the first, showing renewed zeal (Kaiser, 1974).

Temporary territoriality of this type grades into nonterritorial behaviour in which males search along the water for females but have no site attachment although they are aggressive towards other males. *Cordulegaster boltonii* patrols along many hundreds of metres of stream, sometimes turning back and flying over the same part again, and acting aggressively when other males are sighted. Kaiser (1982) has suggested that such aggression serves to space males out and that this benefits them individually since females may arrive at any point and competition from other males is reduced.

When there are too many males for the habitat, some males may behave differently and act as satellites or wanderers. Satellites sit unobtrusively close to or within a male's territory being either unnoticed or tolerated by the resident, and occasionally gaining a female when the resident is occupied. In the American libellulid *Plathemis lydia,* several subordinate males may be tolerated within a dominant male's territory (Campanella & Wolf, 1974). Alternatively males may act as wanderers, passing through the territories of other males or flying elsewhere with no site attachment, occasionally capturing females on their way to the water or when a resident's attention is diverted. Sometimes young males with immature colouring act in this way. In southern Britain, rivers in July or August may support large populations of *Calopteryx splendens* in which only a minority of the males succeeds in holding territories. The majority joins large bands of males which perch near territory holders and pursue females whenever the opportunity arises. Females, however, usually reject such non-territorial males, preferring to accept a resident male after courtship and inspection of the oviposition site. In general, territory holders obtain many more matings than satellites or wanderers, some of which may never mate.

A good way to examine territorial behaviour is to mark and release individual males on their territories (see chapter 6). This is best done when the population density is not too high so that marked males can easily be observed.

Interesting experiments can be done by removing some or all of the males from a small habitat (e.g. a pond) and then examining the behaviour of the remaining few males and of the females. The latter may now outnumber the males and be able to oviposit without interruption. If the captured males are kept in a dark, cool cage, they can be released unharmed after a few hours at the end of the observations (Jacobs, 1955).

Alternatively a male can be removed temporarily from his territory until another takes over, and he can then be released. The ensuing disputes about ownership make an interesting study and the outcome may depend on the time for which the original male was imprisoned. Such experiments can conveniently be carried out with *Calopteryx.*

When water plants form the centre of a territory they can be cut loose and tethered by strings, and thus brought closer to or further away from other territories; the consequent behaviour of residents can then be observed.

Again this can readily be done with *Calopteryx*.

Observations can also be made on how territorial behaviour varies with the time of day, with the season and with the weather. Our knowledge of the behaviour of some British species is still scanty. Further ideas on the study of territoriality and other forms of reproductive behaviour can be found in the chapter by Alcock in Matthews & Matthews (1982).

Recognition

Dragonflies need to be able to identify the species and sex of other individuals. In many species males identify females visually when they approach the water. Females may also need to identify males visually, particularly in genera such as *Calopteryx* in which the male attempts to attract females to his territory. Identification may also depend on tactile mechanisms, mainly by the female, once the pair is in tandem.

In species with brightly coloured males, females can readily be distinguished from rival males. Pajunen (1962) showed by filming that the aggressive approaches of male *Leucorrhinia dubia* to other males were different from their sexual approaches to females, and this has been confirmed in some other species. When the colour of the two sexes is similar, however, identification must depend on other features, one of which may be the characteristic flight movements made by females, particularly during oviposition. Ubukata (1983) attempted to mimic the flight movements of an ovipositing *Cordulia aenea* by gluing a female to a wire on the end of a stick and moving her by hand (fig. 21). He found that simulated oviposition behaviour attracted males. He also glued rolls of paper around a tethered male's abdomen to resemble the

Fig. 21. A female corduliid glued to a wire and moved in a way which attracts the attention of a male. (After Ubukata, 1983.)

thicker abdomen of a female, and concluded that abdominal thickness was another clue used by males. Identification also depended on the response of an individual to an approach by a male: females usually fled whereas other males turned to attack. Ubukata thus concluded that a male identified a female by her movements, her responses and her shape.

Another technique used to explore sexual behaviour is to attach a female to a fishing line so that she can fly but not escape. Either dead or living females may be attached, and the method has been used to examine males' responses to flying or motionless females.

Identification by tactile means is important once a pair is in tandem. There is considerable diversity of form in the males' claspers and, like the genitalia, these have been used by taxonomists to distinguish species. Robertson & Paterson (1982) showed that in female *Enallagma* the pattern of tactile sense organs on the thorax matches the shape of the male's claspers and that this allows the female to identify a male of her species.

There are numerous reports in the literature of pairs in tandem, or even in copulation, involving individuals of two species or even of two different genera. Males may persist in making attempts to copulate but females of the wrong species do not normally co-operate. In flight such tandem pairs sometimes describe an erratic zigzag course, as though the female is attempting to escape. I have seen male *Sympetrum depressiusculum* sit in tandem with immature *Orthetrum cancellatum* of either sex for 2–3 hours, and a male *Ischnura elegans* in tandem and attempting to copulate with a dead male of the same species for over 2 hours. Such prolonged efforts are unusual, however, and a male normally releases the wrong partner after a few minutes. Even three insects in tandem (male–male–female) are not uncommon, and four joined in this way have been reported. Thus it is the female that uses tactile means to identify her partner at close range, and the male shows less discrimination.

Intraspecific visual signals

The most common visual signals used by dragonflies are made by raising or lowering the abdomen. Sometimes the whole abdomen is moved but more commonly only the end segments are used (fig. 22a,b). Male *Calopteryx*, for example, raise these segments to reveal their coloured ventral surfaces before and during courtship. Some libellulids turn their brightly coloured abdomens upwards when they approach rivals aggressively, and other species lower the abdomen.

Fig. 22. Signalling in dragonflies. (a) A male *Calopteryx* reveals the bright patches on the underside of the abdomen. (After Heymer, 1973.) (b) A male *Coenagrion* bobs the abdomen up and down. (After Utzeri and others, 1983.) (c) A female *Anax* in flight turns the abdomen downwards and rejects a male's advance.

(a)

(b)

(c)

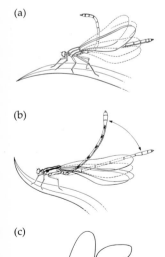

PLATE 1

Rare species
Anisoptera

1
Aeshna isosceles
female

2
Gomphus vulgatissimus
male

3
Leucorrhinia dubia
male

4
Libellula fulva
male

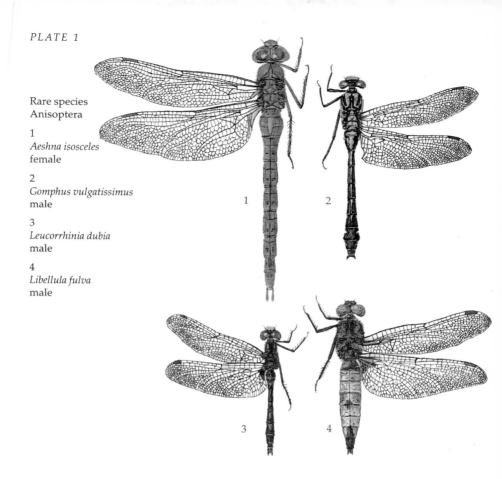

Zygoptera

5
Ischnura pumilio
male

6
Coenagrion mercuriale
male

(Anisoptera life size,
Zygoptera ×1½)

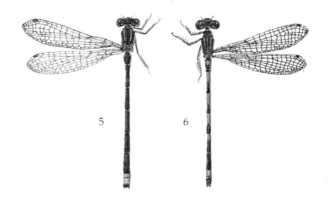

PLATE 2

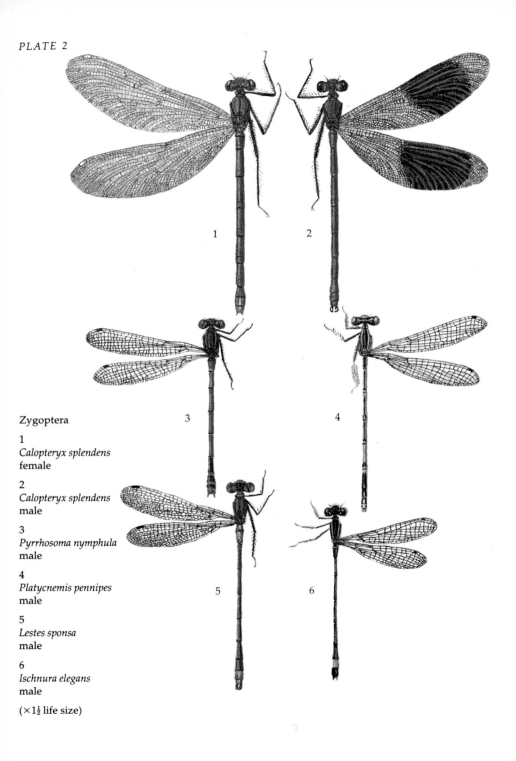

Zygoptera

1
Calopteryx splendens
female

2
Calopteryx splendens
male

3
Pyrrhosoma nymphula
male

4
Platycnemis pennipes
male

5
Lestes sponsa
male

6
Ischnura elegans
male

(×1½ life size)

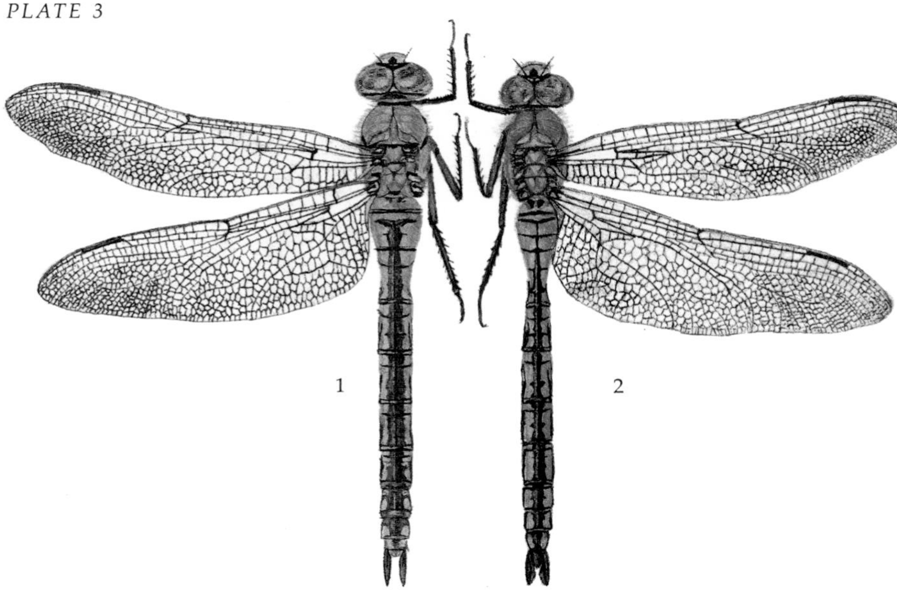

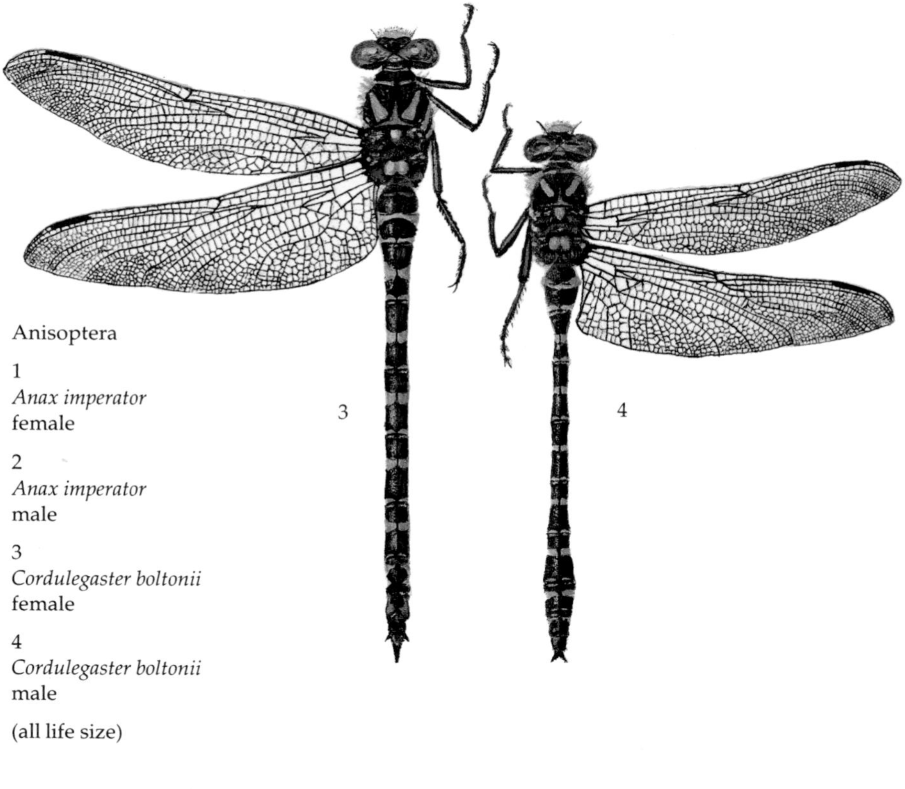

PLATE 3

Anisoptera

1
Anax imperator
female

2
Anax imperator
male

3
Cordulegaster boltonii
female

4
Cordulegaster boltonii
male

(all life size)

PLATE 4

Anisoptera

1
Libellula depressa
female

2
Libellula depressa
male

3
Libellula quadrimaculata
male

4
Sympetrum striolatum
male

5
Cordulia aenea
male

6
Aeshna juncea
male

(all life size)

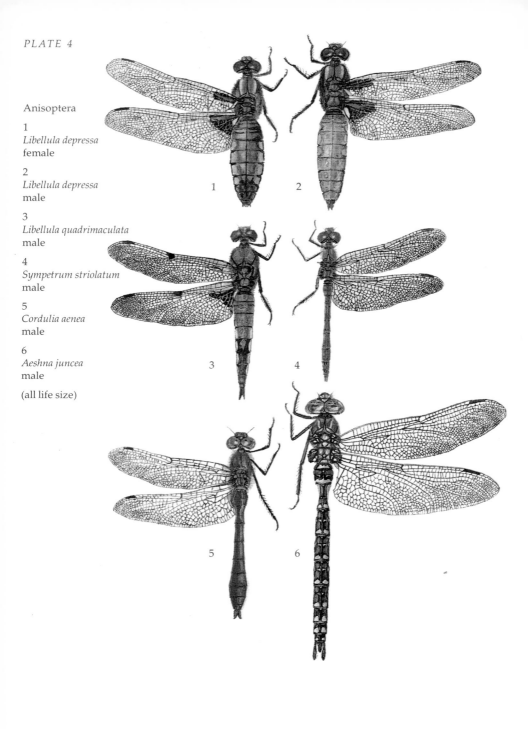

A female may also lower the abdomen or flutter her wings as a rejection signal to a male. For example an ovipositing *Aeshna grandis* flies up about a metre when disturbed by a male and bends her abdomen downwards through 180° forming a hook before resuming oviposition (Robert, 1958), as does *Anax* (fig. 22c). *Ischnura elegans* turns down the end of the abdomen so that it resembles a hockey stick and thereby rejects males when ovipositing.

In some Zygoptera the abdomen may be bobbed up and down repeatedly: this seems to have a wing-cleaning function in many species, but it also acts as a signal in others where it is performed without contacting the wings. Other commonly used signals are wing flaring and wing clapping. Copulating pairs of dragonflies may deter others from approaching by opening and closing their wings several times, and the male may also flex his abdomen, lifting the female upwards. Wing claps, often seen in male *Calopteryx* after settling in their territories, may be used to advertise residency.

Courtship

Although courtship is quite widespread in insects generally, it is uncommon in dragonflies. Courtship is the close-up interaction of male and female which precedes copulation in some species, and it may provide the female with an opportunity to assess the quality of the male. Among dragonflies it has been extensively examined only in members of the Calopterygidae, but it is also known to occur in a few other Zygoptera and in a small number of African and American libellulids.

Male *Calopteryx* perch in their territories and attempt to attract females by displaying the brightly coloured 'tail lights' at the end of the abdomen. When a female perches nearby the male hovers with a type of wing beat, seen at no other time, in which the strokes are shallow and fast, and in some species involve only one pair of wings. He also may land on the water surface, drift downstream and then fly back towards the female. These actions are performed close to the oviposition sites (plants at the water surface) in his territory and may serve to draw the female's attention to them (Heymer, 1973). If the female is receptive she remains perched and the male then lands on her wings and forms a tandem with her.

In Europe *Calopteryx* forms a cluster of closely related species and subspecies in which it would be very interesting to explore further the differences in courtship and other aspects of sexual behaviour which have evolved (Heymer, 1973).

Fig. 23. Copulation in a pair of aeshnids.

Fig. 24. Copulation in a pair of zygopterans.

Copulation and sperm competition

After forming a tandem with a female, male zygopterans settle and then invite copulation by wing flapping and flexing the abdomen. A receptive female responds by bending up her abdomen to the secondary genitalia of the male and they assume the wheel position (fig. 24). Anisopteran copulation normally starts in flight, the male swinging up the female, but again her co-operation is essential and probably no female dragonfly can be forced to copulate. Nevertheless since females cannot escape from the tandem clasp, acceptance of copulation may be the quickest way of obtaining release.

Zygoptera copulate while perched, occasionally flying *in copula* to a fresh perch. Anisoptera may copulate entirely on the wing or they may continue after settling (fig. 23). Aerial copulations may last for only a few seconds, as in *Libellula quadrimaculata*, or persist for 1–2 minutes. Settled copulations usually last longer – in *Sympetrum* for 5–10 minutes, but in *Aeshna cyanea* for up to 2 hours. *Crocothemis erythraea* is unusual in that it occasionally settles for only a few seconds to complete copulation, while *Orthetrum cancellatum*, when active at water, copulates for about 15 seconds, usually completing the process on the ground. However, when away from water, this species may copulate for up to 16 minutes (Siva-Jothy, 1986). Some Zygoptera copulate for about 20 minutes or longer, but species of *Calopteryx* typically take 1–5 minutes. *Ischnura elegans* holds the record for length of copulation time, sometimes copulating for as long as 7–8 hours.

A pair of dragonflies in copulation is vulnerable to predation. Moreover, extended copulations reduce a male's opportunities of mating with other females, and a female's time for oviposition. Selection would therefore be expected to favour rapid mating. Nevertheless copulation is variable in duration in many species and very prolonged in a few. It has been suggested that extended copulations permit a male to transfer more sperm, or that they allow a male to retain a female until she is ready to oviposit, preventing other males from obtaining her, or thirdly that they give more time for sperm competition. Sperm competition provides means whereby males can compete with each other after copulation to fertilise eggs. Waage (1979, 1984) made the remarkable discovery that a copulating male *Calopteryx maculata* removes nearly all the sperm of rivals from a female before he transfers his own sperm to her. Such a female lays eggs fertilised by the last male to have mated with her. Moreover during oviposition the male guards her from other males, ensur-

Fig. 25. The head of the penis of *Orthetrum*. (*a*) Uninflated, (*b*) inflated, (*c*) the end of the flagellum showing the barbs.

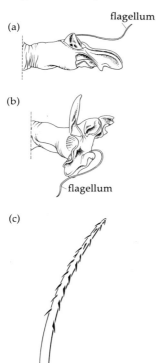

(a)

flagellum

(b)

flagellum

(c)

ing that his sperm are not in turn removed. This type of sperm competition is now known in several other zygopterans and evidence suggests that it occurs in some libellulids as well. Others, however, may merely place their own sperm in the most advantageous position in the female so that it is used first. They may first push their rivals' sperm deep into the female's storage organs before introducing their own. This type of last-in-first-out advantage is common in many insects, whereas active sperm removal using the penis is known only in dragonflies.

By treating males with X-rays it has been confirmed that the last male to mate does fertilise almost all the eggs laid in the next batch (Fincke, 1984; McVey & Smittle, 1984). X-rays sterilise the sperm, and the eggs in consequence are infertile and fail to develop. Within 24 hours after copulation, however, sterile sperm may have mixed with other healthy sperm already in the female and the proportion of infertile to fertile eggs is then the same as that of sterile to healthy sperm.

Examining the structure of the penis may help us to understand how sperm competition is achieved. Those species which remove sperm have a variety of barbs and hooks on structures capable of entering the female's storage organs (e.g. in *Calopteryx* and *Orthetrum*) (fig. 25), while those which copulate rapidly, perhaps packing down rivals' sperm, tend to have blunt penes with large inflatable sacs (e.g. in some species of *Libellula*).

Copulatory activity in Zygoptera may be divisible into two or three stages. In *Enallagma cyathigerum*, for example, the second abdominal segment of a male can be seen to make a series of rapid jerky movements in the first stage, which may persist for 20 minutes, and during which his sperm store remains filled. In the second stage, which lasts for only about 2 minutes, the abdominal position alters and the movement changes to a slower rocking flexure of the third segment which may squeeze his sperm store, emptying it of sperm (fig. 26). During a final short stage the pair remain *in copula* without movement. In the first stage, rivals' sperm is probably removed while in the second the male's own sperm is transferred. The first stage in *Ischnura elegans* sometimes lasts for several hours and the second for over an hour. In this species there are long intervening pauses with no movement, and the prolonged copulation may be partly a means whereby a male retains a female until she is ready to oviposit.

Although females may obtain enough sperm from one copulation to fertilise many batches of eggs, they commonly copulate with several males. For example one female *Crocothemis erythraea* may copulate with up to six males during a single visit of a few minutes to a pond. It has been suggested that females exchange copulations for entry to oviposition sites and for being guarded from other males while ovipositing, and that this explains their acceptance of several males.

The study of copulation and sperm competition provides many opportunities for field and laboratory work. Waage examined sperm removal in *Calopteryx* by measuring the volume of sperm stored in different females caught before, during or after copulation. In this way he found similar volumes before and after, but a much reduced volume at some stages during copulation. Females preserved in alcohol can readily be dissected under a good binocular dissecting microscope by removing the dorsal plates from the eighth and ninth segments.

Fig. 26. Stages in the copulation of *Enallagma cyathigerum*. (*a*) Stage I, jerking. (*b*) Stage II, rocking.

(a)

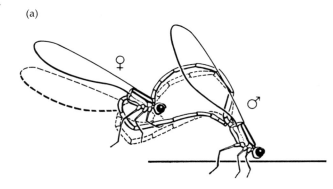

(b)

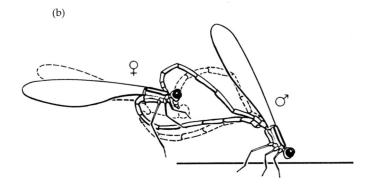

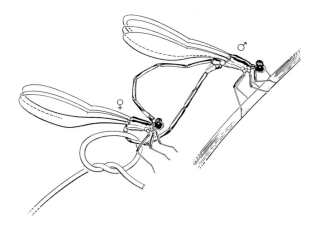

Fig. 27. Copulation in *Ischnura elegans*. The pair has been tethered by a rush stem threaded between male and female, and knotted.

When the gut is pulled away, the sperm-filled storage organs stand out as white structures. Measurement of their volumes is difficult, but a simple alternative is to estimate whether they are full, two-thirds full, one-third full, or empty. Note should also be taken of whether the female contains mature eggs. Since this method requires the killing of several females, it should be used only on very abundant species.

In the field there is still a need for accurate descriptions and measurements of copulatory activity. In Zygoptera it is easy to identify the types of movement occurring, and their duration can be timed. Copulating pairs can be observed through a field monocular and abdominal positions recorded with quick sketches or with a camera; if a video recorder can be used the movements can then be analysed later. Aeshnids and gomphids sometimes copulate in trees and may be difficult to observe, but some libellulids and zygopterans do so low on vegetation or on the ground and are easier to approach. Notes should be made of the time and place of copulation and whether it is in the sun or shade. *Ischnura elegans* is particularly suitable for study because although the pair may be wary, they can easily be caught in a net without interrupting the wheel position, and can then be studied close at hand. Alternatively a wire or a rush stem can be threaded through the pair and used to lift them off their perch for closer study. It should be knotted to prevent their escape (fig. 27), and, remarkably, they will continue to copulate after being securely tethered in this way. They may make intermittent attempts to escape and copulation is usually accelerated under these conditions, which is convenient for the observer.

Fig. 28. A male zygopteran in tandem with a female during oviposition.

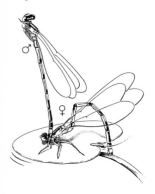

Guarding

The males of many dragonflies guard the females with which they have just mated from the approaches of other males. This usually extends throughout the ensuing bout of oviposition and helps to guarantee the male's paternity of the egg batch. Many zygopterans and some libellulids (e.g. *Sympetrum*) guard ovipositing females by remaining in tandem with them (fig. 28). Moore (1952) found that oviposition movements were made by a male *Sympetrum* when carrying a dead female, and this suggested that the male normally controls the behaviour. In *Enallagma cyathigerum*, in which the female commonly oviposits under water (see below), the male releases the female when she starts to descend and then remains guarding her point of descent. A resurfacing female may then be rescued by the male. In *Erythromma najas* and *Lestes sponsa*, however, both male and female submerge together and they may remain in tandem during a sequence of submergences.

Aeshnids normally oviposit unguarded, but in a few, such as *Aeshna affinis*, *Anax parthenope* and *Hemianax ephippiger*, all of which can be found in southern Europe, the male commonly remains in tandem with the female. He not only guards her from other males, but also serves as a useful lookout when the female, for example, climbs down into hoofprints in mud to oviposit, as has been reported in the first and last of these species. Moreover the male remains ready for flight during cool periods by wing-whirring and is thus able to lift the non-flying female to the next oviposition site. Females which oviposit alone must themselves wing-whirr when it is cool. Thus the female derives several advantages from maintained contact with the male.

Fig. 29. A male libellulid (above) guarding an ovipositing female (on the wing) by perching nearby.

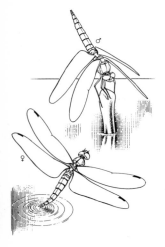

Male *Sympetrum depressiusculum* form tandem pairs with females at dawn, as already described. They remain perched in tandem for at least $2\frac{1}{2}$ hours before copulation and oviposition take place. Such pre-copulatory guarding has also been found in *Coenagrion lindeni* (Utzeri and others, 1983) and other Zygoptera.

Calopterygids and many libellulids guard ovipositing females by hovering or perching close to them ready to drive off intruding males (fig. 29). This method allows them to continue the defence of their territory and to be ready to encounter new females, but it does not guarantee the defence of the first female. Sometimes male *Calopteryx* are found to be guarding more than one female even when they have not mated with them all. Females tend to be attracted to oviposit by the sight of other females ovipositing, and males probably cannot distinguish females they have mated with from others.

Fig. 30. Oviposition into
plants. (*a*) A zygopteran,
(*b*) an aeshnid.

(a)

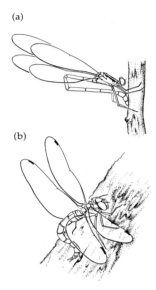

(b)

Oviposition

Three distinct modes of oviposition, termed endophytic, exophytic and epiphytic, can be recognised. In Zygoptera and Aeshnidae, oviposition is nearly always endophytic, which means that the female places eggs within plant tissue or sometimes into mud, using a well-developed ovipositor (fig. 30). Such eggs may be protected from predators and parasites, but oviposition is slow and females are themselves easily caught by predators. Some species insert eggs into plant tissue above the water level, and the eggs may hatch when the water level rises or when the plant falls into the water.

In those zygopterans which descend below the water surface to oviposit, the female, usually alone, clambers about on submerged vegetation inserting her ovipositor into suitable stems. *Enallagma cyathigerum* may stay down for as long as 90 minutes (Doerksen, 1980), but *Calopteryx virgo* holds the record of 123 minutes (Miyakawa, 1982). *Calopteryx splendens* and some other species oviposit either at the surface or after submergence; the factors which determine which she does are unknown, but an investigation of this in the field might be very rewarding. Submerged females look silvery because they carry a layer of air over the abdomen and thorax, and between the wings. This probably acts as a physical gill and allows them to absorb oxygen from the water (Mill, 1974). However, physical gills have a limited life and the duration of the descent must depend partly on the temperature, the activity of the female and the concentration of dissolved oxygen in the water.

Aeshnids oviposit into upright or horizontal plants at or above water level, or sometimes into tree roots, branches or posts. Robert (1958) has even found *Aeshna cyanea* ovipositing into his shoes and his trousers. *Cordulegaster* females thrust the ovipositor down among the pebbles of streams by flying backwards.

Many libellulids oviposit exophytically, scattering clusters of eggs into the water while on the wing. *Sympetrum sanguineum* may drop eggs one at a time onto vegetation near the margin of small ponds. In epiphytic oviposition eggs are placed on the surface of plants: some libellulids hover over partly submerged vegetation tapping their abdomen repeatedly onto the leaves or stems and each time releasing a small cluster of eggs which stick to the plants.

Careful observation in the field allows you to collect the plants on which oviposition has occurred and they can then be searched for eggs. It is also well worth searching for submerged female *Enallagma cyathigerum* where the species is abundant. The eggs may be concealed within the plant tissue but usually the patterns of incisions which are characteristic for a species can be seen (fig. 2). Alternatively it is sometimes possible to persuade females to oviposit after their capture if they are presented with suitable conditions. Thus *Aeshna* will sometimes oviposit into appropriate plants in a cage, and *Enallagma* females will climb down among aquatic vegetation if placed in an aquarium. Under these conditions the oviposition sites can be identified accurately: experiments can be done on the choice of site (e.g. if different plant species are offered), and the rate of oviposition can be measured. In general endophytic species lay at low rates of between 1 and 10 eggs per minute.

Although it is difficult to collect the eggs of exophytic species in the wild, if a gravid female is caught and then her abdomen is dipped into a tube of water she may release clusters of eggs. By rotating the tube the eggs can be made to stick to the walls, and the tube can then be kept until the eggs hatch, after which the young larvae should be placed in an aquarium with small crustaceans, mosquito larvae, etc. for food. Egg release rates have been found by this means to reach 28 per second in some species.

4 Keys to larvae and adults

I Key to larvae

Use of the larval key

The key can be used only to give reliable determinations of last-instar larvae and of the cast skin (exuviae) which remains when the adult has emerged. A larva in the last instar may be readily recognised as the wing cases extend beyond the third abdominal segment; it is often easier to count backwards from the last abdominal segment (the tenth) since the more basal segments can be difficult to distinguish. Most of the characters can be seen with a ×10 lens if the lighting is correctly adjusted, but ideally a stereoscopic microscope with magnifications of (say) ×10 and ×25 or ×30 should be used. It is not necessary to kill a larva to identify it: it should survive the ordeal if it is not left too long under a bright light. With damselfly larvae it is useful to detach one of the gills (lamellae) for examination. This is unlikely to do much harm; healthy specimens are frequently found with one or more lamellae missing. Lamellae are best examined in water with a coverslip and a microscope slide.

The gills of Zygoptera

caudal: pertaining to a tail

lamella: a flat plate

instar: a larval stage between two moults

lateral: at the side(s) of the animal

It is helpful to know a little about the development of the caudal gills (lamellae) of a damselfly larva in order to understand some of the terms used in the key. In the earliest larval stages (instars) the gills are threadlike and hairy. They become triangular in cross-section as development proceeds. In the lateral gills of *Calopteryx* this triangular form is retained. In all other British species the gill becomes flattened in the later instars (fig. I.1). The gill may be a simple complete whole as in the lestids, or there may be traces of a discontinuity running across the gill, recalling a more primitive gill which was made up of two sections jointed together. This discontinuity has been lost in many more advanced forms, but the extent to which it can still be traced is useful in larval taxonomy.

Three types of lamella can be distinguished, apart from the simple gill of lestids and the middle gill of *Calopteryx*:

1. *Nodate lamella*: the node or notch is evidence of the discontinuity or joint which may be traced across the gill. Near to its base, the gill is bordered with many stout setae (hairs). These, the prenodal setae, stop at the node and the gill usually has a rounded tip. E.g. *Erythromma najas*.

I.1

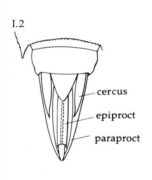

2. *Subnodate lamella*: the position of the node is marked only by the ending of the row of prenodal setae. There is a faint trace of a joint across the gill. The basal part of the gill is thicker than the tip, which is pointed. E.g. *Enallagma cyathigerum*, *Coenagrion* species, *Ischnura* species.

3. *Denodate lamella*: all trace of the node has been lost and there is no apparent difference between the pre- and postnodal setae. E.g. *Pyrrhosoma nymphula*. *Coenagrion mercuriale* also has lamellae which approach this condition.

It can be difficult to decide whether a lamella is subnodate or nodate as the categories tend to merge into one another; this can be a problem in the genus *Coenagrion*. Where this could cause confusion in the key, additional characters have been supplied.

Note: Coenagrion armatum, C. lunulatum and *C. scitulum* are not included.

I.2

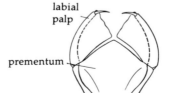

cercus

epiproct

paraproct

tarsal segments: segments on the foot

1 Tip of the abdomen with three appendages which are flat lamellae (I.1) (in all species except *Calopteryx* in which the middle appendage is flat but the two lateral appendages are triangular in section)   (suborder Zygoptera)   25

– Tip of the abdomen with five appendages, none of which is flat (I.2)   (suborder Anisoptera)   2

2 Labial mask flat (I.3)   3

– Labial mask concave and forming a spoon-shaped structure (I.4)   4

3 Abdomen wide and flat; antennae short, thick and 4-segmented (I.5); hind legs with 3 tarsal segments, others with 4   *Gomphus vulgatissimus* (family Gomphidae)

– Abdomen narrow and more or less circular in cross-section; antennae long, threadlike and 7-segmented; all legs with 3 tarsal segments   (family Aeshnidae)   6

I.3

labial palp

prementum

I.4

labial palp

prementum

I.5

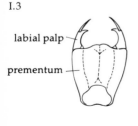

4 Ends of labial palps cut into large irregular teeth, without setae (I.6); larva at least 30 mm long
Cordulegaster boltonii (family Cordulegasteridae)

– Ends of labial palps smooth or scalloped in a shallower, more regular way, with setae (I.7); larva less than 30 mm long 5

I.6

irregular teeth

I.7

setae

I.8

9
10
b *a*
(*b* < 2*a*)

5 Cerci at least half as long as paraprocts (I.8); ends of labial palps more deeply scalloped (family Corduliidae) 13

– Cerci less than half as long as paraprocts (I.9); ends of labial palps shallowly scalloped
(family Libellulidae) 16

I.9

9
10
b *a*
(*b* > 2*a*)

6 Eyes small; sides of head tapering (I.10); appendages at tip of abdomen about equal in length to abdominal segment 10 *Brachytron pratense*

– Eyes larger; sides of head not tapering (I.11); appendages at least twice as long as abdominal segment 10 7

7 Length and width of head about equal, giving an approximately circular appearance (I.11); length at least 50 mm
Anax imperator

– Length of head less than width; length less than 50 mm 8

8 Cerci about $\frac{2}{3}$ length of paraprocts (I.2) *Aeshna isosceles*
– Cerci less than $\frac{1}{2}$ length of paraprocts (I.2) 9

9 Small insect, less than 38 mm long 10
– Large insect, more than 38 mm long 11

I.10

I.11

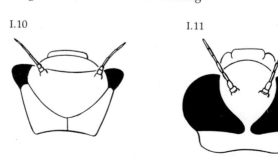

I.12   I.13

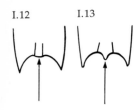

10 Lateral spines on abdominal segment 9 reaching at least
   as far as middle of segment 10; end of epiproct evenly
   hollowed (I.12)   *Aeshna mixta*
 – Lateral spines on abdominal segment 9 shorter, not
   reaching as far as the middle of segment 10; end of epiproct with a small central process interrupting the evenly
   hollowed outline (I.13)   *Aeshna caerulea*

I.14

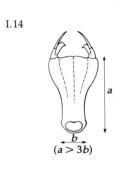

$(a > 3b)$

dorsal: on the animal's back

11 Prementum long and slender; its length more than 3
   times its basal width (I.14)   *Aeshna cyanea*
 – Prementum short and wide; its length about twice its
   basal width (I.15)   12

12 Characteristic marks on thorax and with banded legs
   (I.16); definite lateral spine on segment 6 of abdomen;
   lateral spines on abdominal segment 9 reaching at least to
   middle of segment 10 in most specimens   *Aeshna grandis*
 – No definite markings on legs or thorax; lateral spine on
   segment 6 minute or absent; lateral spines on abdominal
   segment 9 not reaching as far as middle of segment 10
                                              Aeshna juncea

13 Abdomen with prominent spines on the midline above,
   but without prominent dorsal setae (I.17 and I.18)   14
 – Abdomen without dorsal spines, but dorsal surface
   partly covered with prominent setae (I.20 and I.21)   15

I.15

$(a \approx 2b)$

I.17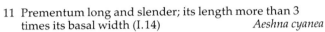

I.18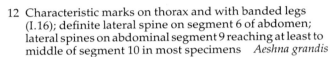

I.20

I.16

I.21

I.19

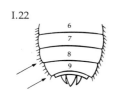

I.22

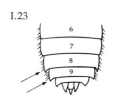

I.23

14 Dorsal spines small, the height of those on segments 6–9 less than half the width of their respective segments (I.17); the spine on segment 9 minute or absent; characteristic dark markings on thorax (I.19)　　*Cordulia aenea*

– Dorsal spines larger, the height of those on segments 6–9 always more than half the width of their respective segments, the spine on segment 9 conspicuous (I.18); sides of thorax uniformly coloured　　*Somatochlora metallica*

15 No lateral spine on segments 8 and 9 (I.22)
　　　　　　　　　　　　　　　　Somatochlora arctica

– Small lateral spines on segments 8 and 9 (I.23)
　　　　　　　　　　　　　　　　Oxygastra curtisii

16 Prominent lateral spines on segments 8 and 9; legs long and slender; tibial–tarsal joint reaching beyond the tip of the abdomen when hind legs are pulled out straight and parallel to the abdomen (I.24)　　　　　　　　21

– Lateral spines on segments 8 and 9 small or absent; legs short and stout; the tibial–tarsal joint not reaching as far as the tip of the abdomen (I.25)　　　　　　　　17

I.24　　　　　　　　　　　　　　　　I.25

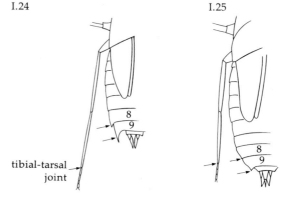

tibial-tarsal joint

I.26

I.27

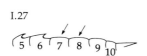

I.28

I.29

17 No dorsal spine on segment 8 (I.26 and I.27)　　　　18

– Dorsal spine present on segment 8 (I.28 and I.29)　　19

18 Length less than 20 mm; sometimes a small dorsal spine on segment 7 (I.26); length of epiproct about 1.5 times its basal width　　　　　　　　*Orthetrum coerulescens*

– Length more than 20 mm; no dorsal spine on segment 7 (I.27); length of epiproct nearly twice its basal width
　　　　　　　　　　　　　　　　Orthetrum cancellatum

I.31

19 Dorsal spines long and curved; a dorsal spine present on segment 9 (I.28) *Libellula fulva*
 – Dorsal spines smaller and less strongly curved, and absent on segment 9 (I.29) 20

20 Labial palps with deep serrations (ratio $l:h$ approx $3:1$) (I.30a) *Libellula depressa*
 – Labial palps with very shallow serrations ($l:h$ approx $5:1$) (I.30b) *Libellula quadrimaculata*

I.32

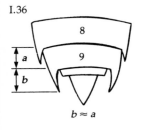

I.33

I.30

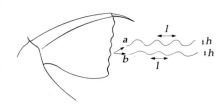

I.34

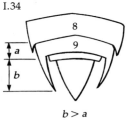

$b > a$

21 Ventral surface of abdomen with conspicuous dark banding (I.31) *Leucorrhinia dubia*
 – Ventral surface without such marking 22

22 Dorsal spine on segment 8 absent or, occasionally, minute (I.32); lateral spines on segment 9 very small, projecting beyond segment 9 by an amount which is less than $\frac{1}{2}$ the length of segment 9 at the level of the lateral spines (I.35) *Sympetrum danae*
 – Pronounced dorsal spine on segment 8 (I.33); lateral spines on segment 9 project beyond segment 9 by more than $\frac{1}{2}$ the length of segment 9 at the level of the lateral spines (I.34 and I.36) 23

I.35

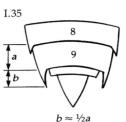

$b \approx \frac{1}{2}a$

23 Lateral spines on segment 9 smaller, projecting beyond the segment by an amount a little less than the length of segment 9 at the level of the lateral spines (I.36) *Sympetrum sanguineum*
 – Lateral spines on segment 9 larger, projecting by more than the length of segment 9 (I.34) 24

I.36

$b \approx a$

24 Lateral spines on segment 9 curved; dorsal spines on segments 5–8 curved, and that on segment 4 absent or minute *Sympetrum striolatum*
 – Lateral spines on segment 9 almost straight; dorsal spines present on segments 4–8 *Sympetrum nigrescens*
[*Note*: the status of *Sympetrum nigrescens* is not clear at present.]

I.37

25 First antennal segment elongated, as long as the combined length of the remaining segments (I.37); middle caudal appendage flat and outer appendages approximately triangular in cross-section
(family Calopterygidae) 26

– First antennal segment very short, shorter than the combined length of the remaining segments; all three caudal appendages flat 27

I.38

26 Front edge of thorax, seen from above, with a strongly pointed process (I.38) *Calopteryx virgo*

– Front edge of thorax with less strongly pointed process (I.39) *Calopteryx splendens*

I.39

27 Labial mask very narrow in its basal half; a cleft in the midline at the end (I.40a) (family Lestidae) 28

– Labial mask not so abruptly narrowed, tapering uniformly towards the base; without a midline cleft (I.40b) 29

I.40 (a) (b)

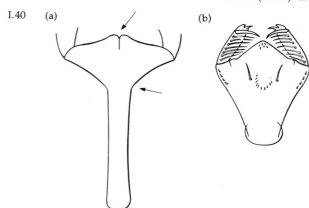

I.41

28 Outer caudal lamellae almost parallel-sided *Lestes sponsa*

– Outer caudal lamellae almost parallel-sided for basal half only, and then tapering; rare species *Lestes dryas*

I.42

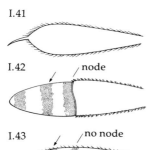

29 Caudal lamellae pointed with threadlike tips (I.41)
Platycnemis pennipes (family Platycnemididae)

– Caudal lamellae rounded or pointed but without threadlike tips (I.42, I.43, I.46–I.48)
(family Coenagrionidae) 30

I.43

30 Lamellae marked with 3 dark bands across them; nodate (I.42) *Erythromma najas*

– Lamellae not marked as in I.42 31

I.44

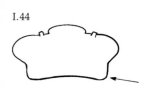

31   Lamellae denodate and usually marked with a dark X-shaped pattern (I.43); head almost rectangular behind (I.44)   *Pyrrhosoma nymphula*

–   Lamellae subnodate or nodate and not marked as in I.43; sides of head converging and not perpendicular to the rear margin (I.45)   32

32   Prementum usually with one seta, rarely 2 setae, on each side (I.40*b*)   *Ceriagrion tenellum*
(Pull mask forwards and view with stereoscopic microscope)

I.45

–   Prementum with at least 3 setae on each side   33

33   Rear of dorsal surface of head strongly spotted (I.45); caudal lamellae nodate   34

–   Rear of dorsal surface of head either without spots or very indistinctly spotted; caudal lamellae subnodate   35

I.46

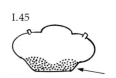

34   Antennae 6-segmented; nodal line running straight across lamellae at right-angles to margins (I.46)   *Coenagrion hastulatum*

–   Antennae 7-segmented; nodal line curved and not perpendicular to margins (I.47)   *Coenagrion puella/pulchellum*

I.47

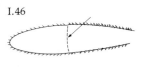

35   Caudal lamellae with the thicker setae reaching to (or beyond) midpoint on only one margin and to about $\frac{1}{4}$ of the length on the other margin (I.48)   36

I.48

–   Caudal lamellae with the thicker setae reaching to (or beyond) midpoint on both margins   37

36   Prementum with 4 or 5 setae on each side; labial palps usually with 6 setae each (I.40*b*); on abdominal segments 7 and 8, the spines on the lateral ridges are stouter than those on the ventral surface   *Ischnura elegans*

–   Prementum with 6 setae on each side; labial palps usually with 5 setae each; on abdominal segments 7 and 8, the spines on the lateral ridges are about the same thickness and length as those on the ventral surface   *Ischnura pumilio*

37   Lamellae short and narrow, approximately parallel-sided for $\frac{1}{2}$ their length; setae much longer and denser after the node than before it; antennae 7-segmented; length (including lamellae) about 16 mm   *Coenagrion mercuriale*

–   Lamellae longer and wider, with margins convex, not parallel-sided; setae before and after the node of approximately the same length and density; antennae 6-segmented; length (including lamellae) about 23 mm   *Enallagma cyathigerum*

II   Key to adult dragonflies

To separate males from females use a hand-lens to look for the secondary reproductive organs of the male on the undersides of abdominal segments 2 and 3 (fig. 17). Female Zygoptera and Aeshnidae are both distinguished by the prominent ovipositor at the end of the abdomen. Lengths given are from head to end of abdomen.

Note: Coenagrion lunulatum is not included.

II.1. Damselfly *Pyrrhosoma nymphula* (male)

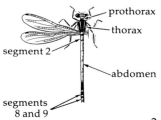

1   Hind wing narrowed in basal half so that its widest part is beyond the mid-point; eyes widely separated and not covering the top of the head (II.1); flight weak and wing beats slow

suborder Zygoptera (damselflies)   2

–   Hind wing widest near base; fore and hind wings different in shape; eyes usually covering much of the head (II.2); flight strong and wing beats rapid

suborder Anisoptera (hawkers, darters and emeralds) 26

II.2 Hawker dragonfly *Anax imperator* (male)

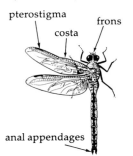

2   Abdomen red, with or without black markings (the rare variety *aurantiaca* of *Ischnura pumilio* should not be confused here as it is orange, not red)   3

–   Abdomen not red   4

3   Legs black   *Pyrrhosoma nymphula* (II.1, pl. 2.3)

–   Legs red/orange   *Ceriagrion tenellum*

Ceriagrion tenellum males occur in only one form with a wholly red abdomen. The females, however, have three forms:
(i) abdomen ⅓ red, ⅔ black: the typical form
(ii) abdomen all black: *melanogastrum*
(iii) abdomen all red (as in the male): *erythrogastrum*. This form is rare and any record of it is of interest.

4   Abdomen pale green or white with black markings; legs whitish and flattened   *Platycnemis pennipes*

(recently emerged individuals and females)

–   Not as above   5

5   Abdomen metallic green or blue   6

–   Not as above   '11

6   Wings coloured or at least yellowish; large insects at least 50 mm long. Male body brilliant metallic blue; wings at least partly blue-black; no pterostigma. Female body metallic green; wings brown to greenish; conspicuous false white pterostigma on wing   7

–   Wings clear; smaller insects less than 40 mm long   8

II.3

7 Male wings almost all blue-black (dark brown in newly
   emerged insects); female wings brownish
                                    Calopteryx virgo

– Male wings with central blue-black band; female wings
   greenish   *Calopteryx splendens* (pl. 2.1, 2.2)
   (The false pterostigma in the female *C. splendens* is nearer the
   wing tip than in *C. virgo*, and the wing venation is denser in
   C. virgo.)

8 Males, with powder-blue markings on base and tip of
   abdomen (unless recently emerged)   9

– Females, with very obvious ovipositor   10

II.4

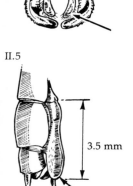

9 Lower anal appendages straight (II.3)
                                    Lestes sponsa (pl. 2.5)

– Lower anal appendages clubbed (II.4)   *Lestes dryas*

10 Ovipositor 3.5 mm long (II.5); upper surface of abdominal
   segment 1 with squarish green markings (II.6)
                                    Lestes dryas

– Ovipositor 2.5 mm long (II.7); upper surface of abdominal
   segment 1 with rounded green markings (II.8)
                                    Lestes sponsa

11 Eyes distinctly red; male abdomen black with sky-blue
   markings at each end   *Erythromma najas*

II.5

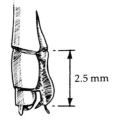

3.5 mm

– Not as above   12

12 Upper surface of abdomen mainly black and with blue
   markings confined to segments 8 and 9   13

– Abdomen either all dark or with an extensive blue pattern
                                                           17

II.7

2.5 mm

II.6

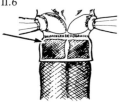

II.8

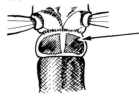

II.9

13 Pterostigma diamond-shaped and two-coloured (II.9) 14
 – Pterostigma rectangular and uniformly dark (II.10) 17

14 Males 15
 – Females 16

II.10

15 Abdominal segment 8 sky-blue; abdominal segment 9 dark (II.11); pterostigma similar in fore and hind wings *Ischnura elegans* (pl. 2.6)
 – Abdominal segment 9 and part of abdominal segment 8 sky-blue (II.12); pterostigma in hind wing much smaller than in fore wing (II.13) *Ischnura pumilio* (pl. 1.5)

II.11

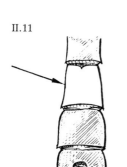

16 Pterostigma same size in all wings *Ischnura elegans*
The females occur in 5 forms:

Form	Thorax colour on dark ground	Abdominal segment 8
Typical	Greenish-blue	Blue
violacea	Violet	Blue
rufescens	Red	Blue
infuscans	Dull brown	Brown
infuscans-obsoleta	Dull brown; no thoracic stripes	Brown

Notes: 1. *violacea* is a young form which matures to either the typical or the *infuscans* form.
  2. *rufescens* is a young form which matures to the *infuscans-obsoleta* form.
  3. Note that the typical form closely resembles the colouring of the male.

II.12

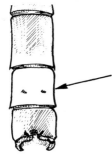

 – Pterostigma in hind wing much smaller than in fore wing (II.13); abdominal segments 8 and 9 same dark colour as other segments *Ischnura pumilio*
The females occur in 2 forms:
  1. Typical: thorax greenish-blue and black; abdomen all black.
  2. *aurantiaca*: thorax orange and black; upper surfaces of abdominal segments 1 and 2 clear orange.

17 Females 18
 – Males 20

II.13 Forewing

II.14

II.15

18 A spine on the underside of abdominal segment 8, at its
hind end (II.14) *Enallagma cyathigerum*

– No such spine (II.15) 19

19 *Coenagrion* species females:
Identification is sometimes difficult as the abdominal
patterning is variable. The shape of the hind margin of
the top of the prothorax, just behind the head, provides
the best means of distinguishing species (II.16*a–e)*

20 End of abdomen mainly black; rest of abdomen with
milky-blue and black pattern; legs whitish and flattened
Platycnemis pennipes (pl. 2.4)

– Not as above 21

II.16. *Coenagrion* species

(a) *C. scitulum*   (b) *C. mercuriale*

(c) *C. hastulatum*   (d) *C. pulchellum*

(e) *C. puella*

21 Abdominal segments 8 and 9 clear blue (II.17*h, i*) 22

– Abdominal segments 8 and 9 with some black markings
(II.17*j–n*) 23

22 Abdominal segment 2 with a black spot on upper surface,
usually stalked (II.17*a*); sometimes an indistinct dark
mark on side of segment 2 *Enallagma cyathigerum*

– Abdominal segment 2 with black mark on upper surface
shaped like a halberd (II.17*b*); clear black mark on each
side of segment 2; eyes green; thorax greenish-blue
Coenagrion hastulatum

23 Anal appendages very conspicuous and longer than last
adominal segment (II.17*n*); abdomen very dark (as in
Ischnura) *Coenagrion armatum*

– Anal appendages inconspicuous 24

24 Abdominal segment 2 with black mark shaped like the sign of Mercury (II.17c)   *Coenagrion mercuriale* (pl. 1.6)

– Not as above (II.17e–g)   25

25 Abdominal segment 2 with U-shaped mark joined at base to a dark line between segments 2 and 3 (II.17f); stripes on top of thorax often broken or reduced to spots
                                                                                          Coenagrion pulchellum

– Abdominal segment 2 with U-shaped mark not joined to line (II.17e); thoracic stripes always complete
                                                                                                Coenagrion puella
(Abdominal markings of *Coenagrion* males are very variable and it is best to check by examining the anal appendages (II.17h–n). The very rare *C. scitulum* can be separated only in this way.)

26 Large hawker dragonflies more than 50 mm long; they tend to make long flights and to perch infrequently   27

– Smaller species (darters or libellulids; club-tailed dragonflies or gomphids; emeralds or corduliids) less than 50 mm long; they tend to make short flights and to perch often but all sometimes make long flights   35

II.17. *Enallagma cyathigerum* and *Coenagrion* species. Upper row: abdominal segment 2 from above (male). Lower row: abdominal segments 9 and 10 from above (male)

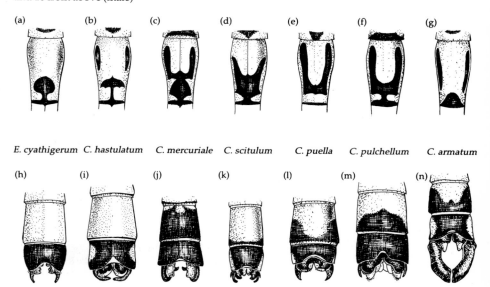

(a)   (b)   (c)   (d)   (e)   (f)   (g)

E. cyathigerum *C. hastulatum*   *C. mercuriale*   *C. scitulum*   *C. puella*   *C. pulchellum*   *C. armatum*

(h)   (i)   (j)   (k)   (l)   (m)   (n)

27 Body black with strong yellow markings; yellow rings
round abdominal segments; large species, male 74 mm,
female 84 mm long　　　*Cordulegaster boltonii* (pl. 3.3, 3.4)

– Not as above　　　　　　　　　　　　　　　　　28

28 Abdomen with a black line running along the top but
otherwise largely blue in mature males and very old
females, or green in most females; thorax green; large
species 78 mm long　　　*Anax imperator* (II.2, pl. 3.1, 3.2)

– Not as above　　　　　　　　　　　　　　　　　29

29 Abdomen brown-black with coloured markings on all
segments　　　　　　　　　　　　　　　　　　　30

– Abdomen uniform brown with few markings　　　　34

30 Pterostigma extremely narrow; abdominal spots pear-
shaped and blue in males but smaller and yellow in
females; small species 55 mm long　　*Brachytron pratense*

– Pterostigma of normal width; abdominal spots not pear-
shaped　　　　　　　　　　　　　　　　　　　31

31 Abdominal segments 9 and 10 crossed by complete
coloured bands (II.22); bands blue in males, yellowish-
green in females; large species 70 mm long　*Aeshna cyanea*

– Abdominal segments 9 and 10 with separate coloured
spots　　　　　　　　　　　　　　　　　　　　32

32 Abdominal segment 2 with cream-coloured T-shaped
mark (II.21); smaller species, 63 mm long　　*Aeshna mixta*

– Abdominal segment 2 without a T-shaped mark　　33

33 Abdominal segments 9 and 10 with notably square-
shaped marks (II.19); thoracic stripes narrow and blue
(greyer in female); abdominal spots all blue (greyer in
some females); costal veins brown; smaller species,
62 mm long　　　　　　　　　　　　　*Aeshna caerulea*

– Abdominal segments 9 and 10 with rounded marks
(II.20); thoracic stripes bold and yellow; abdominal spots
mostly blue in males and green in females; a few yellow
spots always present; costal veins yellow; large species,
74 mm long　　　　　　　　　　　*Aeshna juncea* (pl. 4.6)

34 Wings clear or at most with slight yellowish tint; eyes of
mature insects green; abdominal segment 2 with a cream-
coloured triangle on upper surface (II.18); 67 mm long
　　　　　　　　　　　　　　　　Aeshna isosceles (pl. 1.1)

– Wings tawny-brown; eyes brown and blue; males with
small blue dots on abdomen; large species, 73 mm long
　　　　　　　　　　　　　　　　　　　Aeshna grandis

costal vein: along the front
margin of the wing

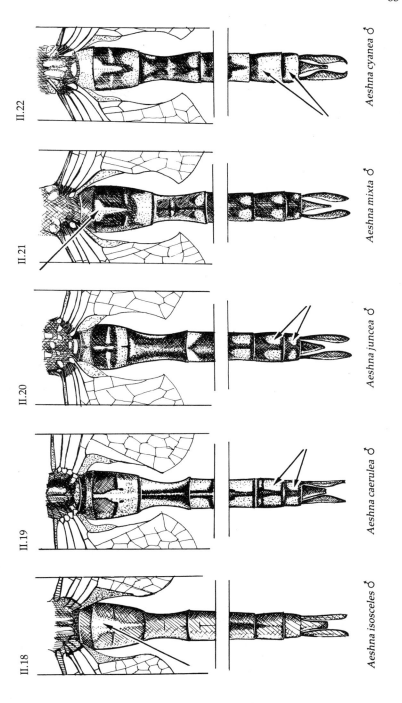

II.22   *Aeshna cyanea* ♂

II.21   *Aeshna mixta* ♂

II.20   *Aeshna juncea* ♂

II.19   *Aeshna caerulea* ♂

II.18   *Aeshna isosceles* ♂

35 Thorax and abdomen black with yellow or green mark-
    ings; eyes well separated at top of head
                              Gomphus vulgatissimus (pl. 1.2)
 – Markings not as above; eyes touching at top of head   36

36 Male abdomen powder-blue   37
 – Male abdomen not blue   40

37 Wings with black marks at bases (II.27)   38
 – Wings clear at bases   39

38 Abdomen conspicuously flattened and broad with small
    lemon-yellow marks along the sides; top of thorax with
    two indistinct blue stripes   *Libellula depressa*
                                             (males) (pl. 4.2)
 – Abdomen narrower and more tapering with a black tip;
    no yellow marks; thorax uniformly dark
                                       Libellula fulva (pl. 1.4)

39 Wings with dark costal veins and large light-brown
    pterostigma; top of thorax with two creamy stripes (indis-
    tinct in old specimens) (II.34); small species, 42 mm long
                                       Orthetrum coerulescens
 – Wings with light costal veins and small black pterostigma
    (II.32); larger species, 50 mm long   *Orthetrum cancellatum*

40 Abdomen and thorax completely black; very small
    species, 32 mm long   *Sympetrum danae*
                                             (males) (pl. 1.3)
 – Not as above   41

41 Abdomen and thorax black with red markings; wing
    bases black; front of head white; 37 mm *Leucorrhinia dubia*
                                             (males)
 – Not as above   42

42 Body metallic green (looking very dark in flight)
    (emeralds)   43
 – Body not metallic green (usually red or tawny)   44

43 Top of abdomen with yellow or orange spots (II.23)
                                             Oxygastra curtisii

                                        ⎧ *Cordulia aenea* (pl. 4.5)
 – No such markings on abdomen   ⎨ *Somatochlora metallica*
                                        ⎩ *Somatochlora arctica*
(These species are readily distinguished by their anal appendages
(II.24, II.25))

II.26

II.27

44 Wings with black bases (II.27) 45
 – Wing without black bases 48

45 Small species, 37 mm long; front of head white; abdomen
   and thorax black with yellow markings *Leucorrhinia dubia*
   (recently emerged individuals and females)
 – Larger tawny-coloured species 46

46 Costal vein on each wing with a dark spot at the centre
   (II.26); abdomen tawny, darkening towards the tip and
   with lemon yellow marks along the sides (indistinct in
   older insects) *Libellula quadrimaculata* (pl. 4.3)
 – Costal vein with no dark spot 47

47 Wings darkened at tip; abdomen tawny with thick black
   stripe along the top widening at the hind end (II.27)
   Libellula fulva
   (recently emerged individuals and females)
 – Wings clear; abdomen conspicuously flattened and
   broad, with bright lemon-yellow marks along the sides
   Libellula depressa
   females (pl. 4.1)

48 Abdomen red (*Sympetrum* species males) 49
 – Abdomen not red, usually tawny or brown 54

49 Abdomen club-shaped (i.e. broadest part in segments 7
   and 8) and blood-red in males; legs black; small species,
   34 mm long *Sympetrum sanguineum*
 – Abdomen not club-shaped; legs with at least a yellow
   stripe along part of their length 50

II.24. End of abdomen from
side (female) showing vulvar
scales

(a) *Somatochlora arctica*

(b) *Somatochlora metallica*

II.25. Male anal appendages

(a) *Cordulia
aenea* (b) *Somatochlora
metallica* (c) *Somatochlora
arctica*

66 Key to adults

50 Hind wings with yellow patch covering basal third
Sympetrum flaveolum
 – Wings with at most a small patch of yellow at base 51

51 Wings with bright scarlet veins in front portion
Sympetrum fonscolombei
 – Wings with brown or dark-red, but never bright scarlet, veins 52

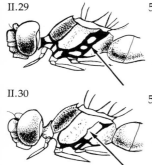

52 Front of head with dark band extending across the top but not down the sides (II.28); pattern on sides of thorax as in II.29 *Sympetrum striolatum* (pl. 4.4)
 – Front of head with dark band extending across top and at least partly down the sides 53

53 Front of head with black band extending slightly down the sides; pattern on sides of thorax as in II.30
Sympetrum nigrescens
(status in doubt)
 – Front of head with black band extending fully down each side (II.31) *Sympetrum vulgatum*

54 Abdomen with two longitudinal stripes along the top (II.32); large species, 50 mm long *Orthetrum cancellatum*
(recently emerged individuals and females)
 – Markings not as above; smaller species, not more than 40 mm long 55

II.32

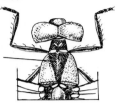

II.33

II.34

55 Top of thorax in front of wings with triangular black mark (II.33); sides of thorax with clear black and yellow marks; very small species, 32 mm long   *Sympetrum danae*

–   Markings not as above; larger species, at least 35 mm long
                                                                                          56

56 Top of thorax in front of wings with two distinctive cream-coloured stripes (II.34); thorax with dark ground colour (not yellowish)   *Orthetrum coerulescens*
                                       (recently emerged individuals and females)

–   Top of thorax with stripes very indistinct or absent; thorax with yellow-brown ground colour   57

57 Legs wholly black; small species, 35 mm long
                                                                        Sympetrum sanguineum
                                       (recently emerged individuals or females)

–   Legs with at least a yellow stripe along part of their length
                                                                                 return to 49
    (Recently emerged individuals and females of other *Sympetrum* species can be identified using the same features as for males from this point onwards.)

5 Conservation and the dragonfly recording scheme

The dragonfly fauna of Britain is dwindling and 11 of our 39 breeding species are under threat or exist precariously, as already indicated (Moore, 1976). Threats to dragonflies arise primarily from the drainage and reclamation of wasteland and from pollution which may result from various agricultural, industrial or sewage disposal practices. At present the government spends only a small fraction of the money on nature conservation that it spends on agricultural development. Collectors seldom threaten dragonfly species although they can occasionally affect small localised populations, and collectors should always take only the minimum number of specimens.

Wetlands such as bogs, marshes, swamps, ponds, streams, rivers and lakes provide rich and varied habitats for many animals and plants, and their preservation is usually compatible with angling, sailing and other sports. In 1950, 32 species of dragonfly bred in national nature reserves and efforts have been made to increase their abundance by creating fresh habitats. For example 20 small ponds were dug in Woodwalton Fen Reserve, Huntingdonshire, and about 11 species now breed in them. Bomb craters have in the past provided good dragonfly habitats, and in 1984 the army fired explosives at a number of sites in its reserves in Dorset, with the aim of making ponds for dragonflies. A better use for such military skills is hard to imagine, and the military interest in conservation deserves every encouragement since many fine stretches of countryside are under military control.

However, creating new habitats or designating existing ones as Sites of Special Scientific Interest is not always enough since drainage or pollution in the neighbourhood can have widespread effects. For example *Oxygastra curtisii* became extinct in Britain about 30 years ago when a new sewage works upstream from its protected habitat polluted the stream (Moore, 1976; see Chelmick, 1980).

Active naturalists can play a vital role in conservation. They can ensure that their local County Naturalists' Trusts are fully aware of the need to preserve a variety of aquatic habitats and they can assist local water boards to be vigilant against pollution. Changes in dragonfly populations are sometimes one of the earliest signs of pollution. Cloudiness of the water and an abundance of algae are other signs to look out for.

Good dragonfly habitats should have clear water in

which a variety of submerged aquatic vegetation can flourish. There should also be a good fringe of emergent sedges, reeds or bulrushes and there should be trees in the vicinity but not so close as to overshadow too much of the habitat. Small dragonfly ponds can readily be constructed in nature reserves, schools or in back gardens. They can be quickly scooped out with a JCB, or more laboriously dug by hand. A pond only 2–3 metres in diameter and 0.5 metres deep will suffice. If it needs to be lined, two or three layers of heavy-duty builders' plastic sheeting protected by a layer of inverted turves will serve adequately. The turves provide soil for plants to grow in; there should also be a shallow margin where emergent species can flourish. With such a pond, dragonflies should soon arrive.

Dragonfly recording scheme

During the last few years intensive efforts have been made to bring the dragonfly records of Britain up-to-date. This has been coordinated at the Institute of Terrestrial Ecology's centre near Huntingdon (Monks Wood) where R. Merritt has organised the information and has published it in the second edition of *The Dragonflies of Great Britain* (Hammond, 1983). Records are now available from 1161 of the 2660 10-kilometre grid squares into which Britain is divided. Of these, 395 contain more than 7 species of dragonfly, 29 have more than 20 species, and in Surrey one has 27 species. There is a continual need to keep such information up to date, and records from your own district may be very valuable.

6 Some techniques for studying dragonflies

Whereas it is not worth while attempting to keep adults in captivity, larvae can easily be reared in aquaria. They need plenty of aquatic vegetation and if they are sprawlers or burrowers a good bottom deposit should be added. There should also be sticks to climb out on if emergence is imminent. *Tubifex*, *Daphnia* and chironomid and mosquito larvae can be used to feed the young stages, and earthworms are suitable for later ones. Larvae should be fed every day in warm conditions, but in winter they require less food particularly if kept cool. A common cause of death is stagnation of the water due to dead vegetation or prey, and a bubbler should be used to aerate the water.

Equipment for field work should include a close-focusing field monocular (available from Microinstruments, Little Clarendon Street, Oxford): most ornithological field binoculars do not focus close enough. A note-book and pencil are essential, and it is well worth carrying a small thermometer and a stop watch. A small tape recorder (e.g. Philips 585 pocket recorder) is very handy when you want to record fast events. Dragonflies make excellent photographic subjects and good photos can be obtained with a 35 mm camera fitted with a flash, a 100 mm lens and extension tubes. Video recorders are much more expensive, but when fitted with a telephoto lens are ideal for recording complex behaviour which can then be analysed by running the tape slowly.

A good, large-mesh net with a long handle is best for catching dragonflies, which can then be kept alive in small envelopes stacked in a cardboard box. For marking adults, coloured spots or numbers can be put on the wings with cellulose paint, or better with a waterproof felt pen (e.g. Lumocolor). Marks should be visible from several metres and they should be readable with a field monocular so that an individual dragonfly can be identified without being disturbed. Ideally the mark should last for several weeks and it should cause a minimum of physical or behavioural disturbance to the insect. A marked dragonfly should be released as gently as possible so as not to evoke escape flight. Marking can sometimes be carried out at the roost in the early morning when dragonflies are too cold to fly. Marking at emergence is very valuable because the age of adults can then be known, but immature dragonflies are very easily damaged when caught.

Larvae are difficult to mark, but their surface can be dried sufficiently to be painted. Alternatively tarsal segments can be cut off, or tiny rings of aluminium (cut from a milk-bottle top) placed around a leg. It must be remembered that such marks will be lost after a moult.

In territorial studies of adults it is convenient to place numbered stakes at 1 metre intervals along a bank or in the water. Not only do they provide extra perches, but they also enable a male's position to be recorded rapidly. The number of oviposition sites can sometimes be increased by adding appropriate species of water plants; by moving such plants about on strings, territory position and size can be adjusted, with interesting consequences for the resident males.

Collecting dragonflies for the cabinet should be kept to a minimum. For identification, photographs can sometimes provide an adequate substitute. If a collection is needed, dragonflies can be caught and killed in the vapour of ethyl acetate or by freezing. Their colours can be preserved best if they are freeze-dried. When this is impracticable, killing and leaving them in acetone for 24 hours improves colour preservation but makes them rather brittle. Dissecting out the gut helps preservation in larger species. Alternatively a wet collection, preserved in 90% alcohol or in 2% formaldehyde, can be made; this allows for the subsequent examination of internal organs.

Dissections of living or preserved insects should be carried out in 1% saline or in water with a good binocular dissecting microscope and a spotlamp. In this way reproductive structures can be located and their state of maturity and sperm contents can be examined. The reproductive organs of both sexes are best dissected from the dorsal side, but the secondary genitalia should be inspected from the ventral side.

Full field notes should always be kept. Record sightings of adults and note the time of day, temperature, weather conditions and the type of activity shown. (Unwin (1980) discusses methods of measuring temperatures and other factors in the field.) Even trivial observations may be found to have significance later. Making estimates of the abundance of dragonflies is difficult, but counting the numbers seen along measured lengths of bank, trying not to count individuals twice, is a simple method. More reliable estimates may be obtained using the capture, mark, release and recapture technique which is described by Davis (1983). It is most suitable for Zygoptera.

Finally, remember that no instrument is as good as the human eye. Patient observation is the most instructive and enjoyable way of learning about dragonflies, and although video cameras and electronic recorders sometimes increase the amount we can learn, they are no substitute for the unaided eye. With no apparatus at all many profitable hours can be spent in pleasant surroundings.

How to present your findings

Writing up is an important part of a research project, particularly when the findings are to be communicated to other people. A really thorough, critical investigation that has established new information of general interest may be worth publishing if the animals on which it is based can be identified with certainty. Journals that publish short papers on insect biology include the *Entomologists' Monthly Magazine, Entomologist's Gazette, Bulletin of the Amateur Entomologists' Society*, and, for material with an educational slant, *Journal of Biological Education*. The most relevant journal may be the *Journal of the British Dragonfly Society*. Those unfamiliar with publishing conventions are advised to examine current numbers of these journals to see what sort of thing they publish, and then to write a paper along similar lines, keeping it as short as is consistent with the presentation of enough information to establish the conclusions. It is then time to consult an appropriate expert who can give advice on whether and in what form the material might be published. It is an unbreakable convention of scientific publication that results are reported with scrupulous honesty. Hence it is essential to keep detailed and accurate records throughout the investigation, and to distinguish in the write-up between certainty and probability, and between deduction and speculation. In many cases it will be necessary to apply appropriate statistical techniques to test the significance of the findings. A book such as Parker's (1973) *Introductory Statistics for Biology* will help, but this is an area where expert advice can contribute much to the planning, as well as the analysis, of the work.

Useful addresses

Societas Internationalis Odonatologica (SIO)

Main office
Dr B. Kiauta, PO Box 256, 3720 AG Bilthoven, The Netherlands

National office in the UK
Dr P. J. Mill (National Representative), Department of Pure and Applied Zoology, The University, Leeds LS2 9JT

British Dragonfly Society

Secretary
Roderick Dunn, 4 Peakland View, Darley Dale, Matlock, Derbyshire DE4 3GF

Journal of the British Dragonfly Society
Editor:
Stephen Brooks, 4 Nelson Road, London N8 9RU

Books and reprints on dragonflies

L. Christie, 129 Franciscan Road, Tooting, London SW17 8DZ

Supply agent for entomological equipment

Watkins and Doncaster, Four Throws, Hawkhurst, Kent

Royal Entomological Society of London

The Registrar, 41 Queen's Gate, London SW7 5HU

Dragonfly recording scheme

R. Merritt, 48 Somersby Avenue, Walton, Chesterfield, Derbyshire S42 7LY

Further reading

Finding books

Some of the books and journals listed here will be unavailable in local and school libraries. It is possible to make arrangements to see or borrow such works by seeking permission to visit the library of a local university, or by asking your local public library to borrow the work (or a photocopy of it) for you via the British Library, Lending Division. This may take several weeks. It is important to present your librarian with a reference that is correct in every detail. References are acceptable in the form given here, namely the author's name and date of publication, followed by (for a book) the title and publisher or (for a journal article) the title of the article, the full journal title, the volume number, and the first and last pages of the article.

References

*An asterisk denotes a highly recommended book.

d'Aguilar, J., Dommanget, J.-L. & Préchac, R. (1986). *A guide to the Dragonflies of Britain, Europe and North Africa*, translated by S. Brooks, N. Brooks and T. S. Robertson. London: Collins.

Alexander, D. E. (1984). Unusual phase relationships between the forewings and hindwings in flying dragonflies. *Journal of Experimental Biology*, **109**, 379–83.

Armett-Kibel, C. & Meinertz-hagen, A. (1983). Structural organization of the ommatidia in the ventral compound eye of the dragonfly, *Sympetrum. Journal of Comparative Physiology*, **151**, 285–94.

Asahina, S. (1974). Development of odonatology in the Far East. *Odonatologica*, **3**, 5–12.

Askew, R. R. (1982). Roosting and resting site selection by coen-agrionid damselflies. *Advances in Odonatology*, **1**, 1–8.

Blois, C. & Cloarec, A. (1985). Influence of experience on prey selection by *Anax imperator* larvae (Aeshnidae: Odonata). *Zeitschrift für Tierpsychologie*, **68**, 303–12.

Bodenheimer, F. S. (1951). *Insects as Human Food*. The Hague: W. Junk.

Butler, S. (1982). *Dragonflies of Shropshire and their Distribution*. Caradoc and Severn Valley Field Club.

Caillère, L. (1974). Modalités du declenchement du comporte-ment de capture chez la larve d'*Agrion* (*Calopteryx*) *splendens*. *Zeitschrift für Tierpsychologie*, **35**, 381–402.

Campanella, P. J. & Wolf, L. (1974). Temporal leks as a mating system in a temperate zone dragonfly (Odonata: Anisoptera). I. *Plathemis lydia* (Drury). *Behaviour*, **51**, 49–87.

Campbell, J. M. (1983). *Atlas of Oxfordshire Dragonflies*. (Available from Oxfordshire C.C., Dept of Museum Services, Fletcher's House, Woodstock, Oxon OX7 1SP.)

Carle, F. (1982). Evolution of the Odonate copulatory process. *Odonatologica*, **11**, 271–86.

Chelmick, D. (1980). Proceedings of the first meeting of British dragonfly recorders. *Notulae Odonatologicae*, **1**, 92–5.

Cloarec, A. (1977). Alimentation des larves d'*Anax imperator* dans un milieu naturel (Anisoptera: Aeshnidae). *Odonatologica*, **6**, 227–43.

Coker, S. & Fox, A. (1985). West Wales Dragonflies. (Available from Mountain Books, Mountain, Clarbeston Rd, Haverfordwest, Pembrokeshire SA63 4SG.)

Consiglio, C. (1976). Some observations on the spacing patterns of *Anax imperator* Leach (Anisoptera: Aeshnidae). *Odonatologica*, **5**, 11–14.

Corbet, P. S. (1957). The life history of the Emperor dragonfly, *Anax imperator*.

Journal of Animal Ecology, **26**, 1–69.

*Corbet, P. S. (1962). *A Biology of Dragonflies*. London: H. F. & G. Witherby (reprinted 1983 by E. W. Classey Ltd).

Corbet, P. S. (1980). Biology of Odonata. *Annual Reviews of Entomology*, **25**, 189–217.

Corbet, P. S. (1984). Orientation and reproductive conditions of migrating dragonflies (Anisoptera). *Odonatologica*, **13**, 81–8.

*Corbet, P. S., Longfield, C. & Moore, N. W. (1960). *Dragonflies*. London: Collins (reprinted 1985).

Davis, B. N. K. (1983). *Insects on Nettles*. Naturalists' Handbooks 1. Cambridge University Press.

Doerksen, G. P. (1980). Notes on the reproductive behaviour of *Enallagma cyathigerum* (Charpentier) (Zygoptera: Coenagrionidae). *Odonatologica*, **9**, 293–6.

Dumont, H. J. & Hinnekint, B. O. N. (1973). Mass migration in dragonflies especially in *Libellula quadrimaculata* L.: a review, a new ecological approach and a new hypothesis. *Odonatologica*, **2**, 1–20.

Dunn, R. (1984). *Derbyshire Dragonflies*. Derbyshire Naturalists' Trust.

Etienne, A. S. (1978). Energy versus time-dependent parameters in the determination of a behavioural sequence in the *Aeshna* larva. *Journal of Comparative Physiology*, **127**, 89–96.

Fincke, O. M. (1982). Lifetime mating success in a natural population of the damselfly *Enallagma hageni* (Odonata: Coenagrionidae). *Behavioural Ecology and Sociobiology*, **10**, 293–302.

Fincke, O. M. (1984). Sperm competition in the damselfly *Enallagma hageni* (Walsh) (Odonata: Coenagrionidae): benefits of multiple matings to males and females. *Behavioural Ecology and Sociobiology*, **14**, 235–40.

Fried, C. S. & May, M. L. (1983). Energy expenditure and food intake of territorial male *Pachydiplax longipennis* (Odonata: Libellulidae). *Ecological Entomology*, **8**, 283–92.

Geijskes, D. C. (1975). The dragonfly wing used as a nose plug ornament. *Odonatologica*, **4**, 29–30.

Geijskes, D. C. & van Tol, J. (1983). *Die libellen van Nederland*. Koninklijke Nederlandse Natuurhistorische Vereniging, Hoegwoud, N.H.

*Hammond, C. O. (1983). *The Dragonflies of Great Britain and Ireland*. 2nd edn (revised by R. Merritt). Colchester, Essex: Harley Books. (Paperback edition (1985).)

Harvey, I. F. & Corbet, P. S. (1985). Territorial behaviour of larvae enhances mating success of male dragonflies. *Animal Behaviour*, **33**, 561–5.

Heinrich, B. & Casey, T. M. (1978). Heat transfer in dragonflies: fliers and perchers. *Journal of Experimental Biology*, **74**, 17–36.

Heymer, A. (1973). Beiträge zur Ethologie und Evolution der Calopterygidae. *Journal of Comparative Ethology*, [1973], Supplement 11.

Higashi, K. (1973). Estimation of the food consumption for some species of dragonflies. I. Estimation by observation for the frequency of feeding flights of dragonflies. *Reports of the Ebino Biology Laboratory, Kyushu University*, **1**, 119–29.

Jacobs, M. E. (1955). Studies on territorialism and sexual selection in dragonflies. *Ecology*, **36**, 566–86.

Johnson, C. (1973). Ovarian development and age recognition in the damselfly, *Argia moesta* (Hagen, 1961) (Zygoptera: Coenagrionidae). *Odonatologica*, **2**, 69–81.

Kaiser, H. (1974). Verhaltensgefüge und Temporialverhalten der Libelle *Aeschna cyanea*. *Zeitschrift für Tierpsychologie*, **34**, 398–429.

Kaiser, H. (1982). Do *Cordulegaster* males defend territories? A preliminary investigation of mating strategies in *Cordulegaster boltoni* (Donovan) (Anisoptera: Cordulegasteridae). *Odonatologica*, **11**, 139–52.

Kanou, M. & Shimozawa, T. (1983). The elicitation of the predatory labial strike of dragonfly larvae in response to a purely mechanical stimulus. *Journal of Experimental Biology*, **107**, 391–401.

Kiauta, B. (1965). Notes on the Odonata fauna of some brackish waters of Walcheren Island. *Entomologische Berichten*, **25**, 54–8.

Komnick, H. (1982). The rectum of larval dragonflies as jet engine, respirator, fuel depot and ion pump. *Advances in Odonatology*, **1**, 69–91.

Krüner, U. (1977). Revier- und Fortpflanzungsverhalten von *Orthetrum cancellatum* L. (Anisoptera: Libellulidae). *Odonatologica*, **6**, 263–70.

Longfield, C. (1948). A vast immigration of dragonflies into the south coast of County Cork. *Irish Naturalists' Journal*, **9**, 133–41.

Lucas, W. J. (1930). *The Aquatic (Naiad) Stage of the British Dragonflies*. London: Ray Society.

McGeeney, A. (1986). *A Complete Guide to British Dragonflies*. London: Jonathan Cape.

McVey, M. E. & Smittle, B. Y. (1984). Sperm precedence in the dragonfly *Erythemis simplicicollis*. *Journal of Insect Physiology*, **30**, 619–28.

Matthews, J. R. & Matthews, R. W. (eds.) (1982). *Insect Behavior. A Source Book of Laboratory and Field Exercises*. Colorado: Westview Press.

May, M. L. (1978). Thermal adaptations of dragonflies. *Odonatologica*, **7**, 27–47.

May, M. L. (1981). Allometric analysis of body and wing measurements on male Anisoptera. *Odonatologica*, **10**, 279–91.

Mill, P. J. (1974). Respiration: aquatic insects. In *The Physiology of Insecta*, ed. M. Rockstein, vol. 6, pp. 403–67. New York & London: Academic Press.

Miyakawa, K. (1982). Reproductive behaviour and life span of adult *Calopteryx atrata* Selys and *C. virgo japonica* Selys (Odonata: Zygoptera). *Advances in Odonatology*, **1**, 193–203.

Moore, N. W. (1952). Notes on the oviposition behaviour of the dragonfly *Sympetrum striolatum* Charpentier. *Behaviour*, **4**, 101–3.

Moore, N. W. (1976). The conservation of Odonata in Great Britain. *Odonatologica*, **5**, 37–44.

Münchberg, P. (1982). On the parasitism on the wings of *Sympetrum meridionale* and *S. fonscolombei* Selys by *Arrenurus papillator* (Muell) (Hydrachnella, Acari) and on the specificity of and some hypotheses about parasitism. *Archiv für Hydrobiologie*, **95**, 299–316.

Neville, A. C. (1983). Daily cuticular growth layers and the teneral stage in adult insects: a review. *Journal of Insect Physiology*, **29**, 211–19.

Norberg, R. A. (1972). The pterostigma of insect wings an inertial regulator of wing pitch. *Journal of Comparative Physiology*, **81**, 9–22.

Norberg, R. A. (1975). Hovering flight of the dragonfly *Aeshna juncea* L. Kinematics and aerodynamics. In *Swimming and Flying in Nature*, ed. T. Y. T. Wu, C. J. Brokaw & C. Brennen. New York: Plenum Press.

Pajunen, V. I. (1962). Studies on the population ecology of *Leucorrhinia dubia* v.d. Lind. (Odon., Libellulidae). *Annales Zoologici Societatis Zoologicae Botanicae Fennicae 'Vanamo'*, **24**, 1–79.

Parker, R. E. (1973). *Introductory Statistics for Biology*. Studies in Biology no. 43. London: Edward Arnold.

Parr, M. J. (1976). Some aspects of the population ecology of the damselfly, *Enallagma cyathigerum* (Charpentier) (Zygoptera: Coenagrionidae). *Odonatologica*, **5**, 45–57.

Parr, M. J. (1983). An analysis of territoriality in libellulid dragonflies. *Odonatologica*, **12**, 39–57.

Pfau, H. K. (1971). Struktur und Funktion des Sekundären Kopulationsapparates der Odonaten (Insecta: Palaeoptera). *Zeitschrift für Morphologie der Tiere*, **70**, 281–371.

Pond, C. M. (1973). Initiation of flight and pre-flight behaviour of anisopterous dragonflies *Aeshna* spp. *Journal of Insect Physiology*, **19**, 2225–9.

Pritchard, G. (1965). Prey capture by dragonfly larvae (Odonata: Anisoptera). *Canadian Journal of Zoology*, **43**, 271–89.

Pritchard, G. (1982). Life history strategies and the colonization of North America by the genus *Argia* (Odonata: Coenagrionidae). *Advances in Odonatology*, **1**, 227–41.

Robert, P.-A. (1958). *Les Libellules (Odonates)*. Paris: Delachaux et Niestlé.

Robertson, H. M. & Paterson, H. E. H. (1982). Mate recognition and mechanical isolation in *Enallagma* damselflies (Odonata: Coenagrionidae). *Evolution*, **36**, 243–50.

Rowe, R. J. (1980). Territorial behaviour of a larval dragonfly, *Xanthocnemis zealandica*. *Odonatologica*, **9**, 285–92.

Rudolph, R. (1976). Pre-flight behaviour and the initiation of flight in tethered and unrestrained dragonfly, *Calopteryx splendens*. *Odonatologica*, **5**, 59–64.

Rüppell, G. (1985). Kinematic and behavioural aspects of flight of the male banded Agrion, *Calopteryx* (*Agrion*) L. In *Insect Locomotion*, ed. M. Gewecke & G. Wendler, pp. 195–204. Berlin & Hamburg: Paul Parey.

Simmons, P. J. (1982). The function of insect ocelli. *Trends in the Neurosciences*, **5**, 182.

Siva-Jothy, M. T. (1986). Sperm competition in dragonflies. D.Phil. thesis, University of Oxford.

Stavenga, D. G., Bernard, G. D.,

Chappell, R. L. & Wilson, M. (1979). Insect pupil mechanisms. III. On the pigment migration in dragonfly ocelli. *Journal of Comparative Physiology*, **129**, 199–205.

Tanaka, Y. & Hisada, M. (1980). The hydraulic mechanism of the predatory strike in dragonfly larvae. *Journal of Experimental Biology*, **88**, 1–19.

Thompson, D. J. & Pickup, J. (1984). Feeding rates of Zygoptera larvae within an instar. *Odonatologica*, **13**, 309–15.

Ubukata, H. (1983). An experimental study of sex recognition in *Cordulia aenea amurensis* Selys (Anisoptera: Corduliidae). *Odonatologica*, **12**, 71–81.

Unwin, D. M. (1980). *Microclimate Measurement for Ecologists*. New York & London: Academic Press.

Utzeri, C., Falchetti, E. & Carchini, G. (1983). Reproductive behaviour in *Coenagrion lindeni* (Selys) in central Italy (Zygoptera: Coenagrionidae). *Odonatologica*, **12**, 259–78.

Utzeri, C. & Raffi, R. (1983). Observations on the behaviour of *Aeshna affinis* (Van der Linden) at a dried-up pond (Anisoptera: Aeshnidae). *Odonatologica*, **12**, 141–51.

Veron, J. E. N. (1973). Age determination of adult Odonata. *Odonatologica*, **2**, 21–8.

Vogt, F. D. & Heinrich, B. (1983). Thoracic temperature variations in the onset of flight in dragonflies (Odonata: Anisoptera). *Physiological Zoology*, **56**, 236–41.

Waage, J. K. (1979). Dual function of the damselfly penis: sperm removal and sperm transfer. *Science*, **203**, 916–18.

Waage, J. K. (1983). Sexual selection, ESS theory and insect behavior: some examples from damselflies (Odonata). *Florida Entomologist*, **66**, 19–31.

Waage, J. K. (1984). Sperm competition and the evolution of odonate mating systems. In *Sperm Competition and the Evolution of Animal Mating Systems*, ed. R. L. Smith, pp. 257–90. New York & London: Academic Press.

Weis-Fogh, T. (1967). Respiration and tracheal ventilation in locusts and other flying insects. *Journal of Experimental Biology*, **47**, 561–87.

Welstead, N. & Welstead, T. (1984). *The Dragonflies of the New Forest*. Hampshire and Isle of Wight Naturalists' Trust.

Whalley, P. (1980). *Tupus diluculum* sp. nov. (Protodonata), a giant dragonfly from the Upper Carboniferous of Britain. *Bulletin of the British Museum (Natural History), Geology*, **34**, 285–7.

Winstanley, W. J. (1982). Observations on the Petaluridae (Odonata). *Advances in Odonatology*, **1**, 303–8.

Wojtusiak, J. (1974). A dragonfly migration in the high Hindu Kush (Afghanistan) with a note on high altitude records of *Aeshna juncea* (L.) and *Pantala flavescens* (Fabricius). *Odonatologica*, **3**, 137.

Appendix 1. Check list of British species, with English names and notes on distribution

ORDER ODONATA

Suborder ZYGOPTERA

Family PLATYCNEMIDIDAE
Platycnemis Burmeister, 1839
   pennipes (Pallas, 1771), White-legged damselfly: a southern species reaching only to the level of the Wash. Seldom abundant and very susceptible to pollution.

Family COENAGRIONIDAE
Ceriagrion Selys, 1876
   tenellum (de Villers, 1789), Small red damselfly: southern in distribution but also occurring in north Wales and in a few parts of East Anglia; not usually common.
Erythromma Charpentier, 1840
   najas (Hansemann, 1823), Red-eyed damselfly: mainly southern in distribution, but also found in a few parts of central England.
Coenagrion Kirby, 1890
   armatum (Charpentier, 1840), Norfolk damselfly: used to be found at one locality in Norfolk, but not seen since 1957.
   hastulatum (Charpentier, 1825), Northern damselfly; very rare, occurring only in a few scattered localities in north Scotland.
   lunulatum (Charpentier, 1840), Irish damselfly: very rare, occurring only at a few sites in Ireland.
   mercuriale (Charpentier, 1840), Southern damselfly: very scarce, known only at a few localities in south England and south-west Wales.
   puella (Linnaeus, 1758), Azure damselfly: abundant in England and Wales, but uncommon in Scotland. Widely distributed in Ireland.
   pulchellum (Vander Linden, 1825), Variable damselfly: much less common than *C. puella*, but with a similar distribution.
   scitulum (Rambur, 1842), Dainty damselfly: known only from one locality in Essex, and not seen since 1953.
Enallagma Charpentier, 1840
   cyathigerum (Charpentier, 1840), Common blue damselfly: very widespread throughout Britain.

Pyrrhosoma Charpentier, 1840
    nymphula (Sulzer, 1776), Large red damselfly: wide-
        spread throughout Britain.
Ischnura Charpentier, 1840
    elegans (Vander Linden, 1823), Blue-tailed damselfly:
        widespread throughout Britain but less abundant in
        Scotland. More resistant to pollution than most other
        species and also tolerant of some salinity.
    pumilio (Charpentier, 1825), Scarce blue-tailed damsel-
        fly: scarce, occurring in a few parts of south and
        south-west England and in Wales. It is also known in
        several places in Ireland.

Family LESTIDAE
Lestes Leach, 1815
    dryas Kirby, 1890, Scarce emerald damselfly: very
        scarce; it used to be found at a few localities in East
        Anglia and Essex but not seen recently. Still occurs in
        parts of Ireland.
    sponsa (Hansemann, 1823), Emerald damselfly: wide-
        spread throughout Britain and Ireland, and some-
        times abundant.

Family CALOPTERYGIDAE
Calopteryx Leach, 1815
    splendens (Harris, 1782), Banded demoiselle: common
        in the southern half of England and throughout
        Wales; it also occurs less commonly at occasional
        localities in north England and in Ireland.
    virgo (Linnaeus, 1758), Beautiful demoiselle: locally
        common in south and south-west England and in
        Wales. It also occurs at a few sites in north England,
        west Scotland and in Ireland.

Suborder ANISOPTERA
Family GOMPHIDAE
Gomphus Leach, 1815
    vulgatissimus (Linneaus, 1758), Club-tailed dragonfly:
        quite common in a few regions of the Thames, Severn
        and Wye. Also found at a very few other sites in
        southern England, and in central and south-west
        Wales.

Family AESHNIDAE

Brachytron Selys, 1850

    pratense (Müller, 1764), Hairy dragonfly: widely scattered but never common, occurring in south England, Wales and Ireland.

Aeshna Fabricius, 1775

    caerulea (Ström, 1783), Azure hawker, widely distributed in north and west Scotland, but never common.

    cyanea (Müller, 1764), Southern hawker: widespread in England but becoming less common in the north. Known also from a very few localities in north Scotland.

    grandis (Linnaeus, 1758), Brown hawker: abundant at many sites in south and central England but not west of Somerset. Not known in Scotland and Wales. Widely scattered in Ireland.

    isosceles (Müller, 1767), Norfolk hawker: known only from a very few sites in Norfolk.

    juncea (Linnaeus, 1758), Common hawker: widely distributed throughout Britain and Ireland, commoner in the west and in uplands.

    mixta Latreille, 1805, Migrant hawker: fairly common in south and east England and in south Wales. Rare north of the Wash. It seems to be extending its range.

Anax Leach, 1815

    imperator Leach, 1815, Emperor dragonfly: common in some parts of south England but not found north of the Wash or in north Wales. Not in Ireland.

Family CORDULEGASTERIDAE

Cordulegaster Leach, 1815

    boltonii (Donovan, 1807), Golden-ringed dragonfly: widely distributed in Britain mainly in the west and north; never very abundant.

Family CORDULIIDAE

Cordulia Leach, 1815

    aenea (Linnaeus, 1758), Downy emerald: local, occurring in south England and also at a few scattered sites in Wales, north England and north Scotland.

Somatochlora Selys, 1871

    arctica (Zetterstedt, 1840), Northern emerald: scarce, occurring only in north and west Scotland.

    metallica (Vander Linden, 1825), Brilliant emerald: occurs at a few sites in south and south-east England and in the north of Scotland.

Oxygastra Selys, 1870
  curtisii (Dale, 1834), Orange-spotted emerald: used to
    occur in the New Forest but no recent record and
    probably extinct in Britain.

Family LIBELLULIDAE
Orthetrum Newman, 1833
  cancellatum (Linnaeus, 1758), Black-tailed skimmer:
    widespread in southern England and East Anglia;
    occurs at a few sites in Wales. Also known in a very
    few localities in Ireland.
  coerulescens (Fabricius, 1798), Keeled skimmer: com-
    mon in some areas in south and south-west England,
    Wales and Ireland. It also occurs in a few scattered
    localities in north-west England and west Scotland.
Libellula Linnaeus, 1758
  depressa Linnaeus, 1758, Broad-tailed chaser: wide-
    spread in south and central England and in Wales.
    Sometimes quickly colonises newly dug garden
    ponds.
  fulva Müller, 1764, Scarce chaser: rare, found only at a
    few localities in south and east England.
  quadrimaculata Linnaeus, 1758, Four-spotted chaser:
    very widespread in Britain and Ireland. A common
    migrant.
Sympetrum Newman, 1833
  danae (Sulzer, 1776), Black darter: common in Scotland
    and found at many scattered sites in England, par-
    ticularly in the west. Common in Wales.
  flaveolum (Linnaeus, 1758), Yellow-winged darter:
    scarce, found only at a few sites in south-east and
    central England. Infrequent migrant.
  fonscolombei (Selys, 1840), Red-veined darter: no recent
    records from Britain. A rare immigrant in east Eng-
    land.
  nigrescens Lucas, 1912, Highland darter: replaces *S.
    striolatum* in north and north-west Scotland. The
    validity of this species is in doubt since intermediate
    forms between it and *S. striolatum* are known.
  sanguineum (Müller, 1764), Ruddy darter: widely scat-
    tered in south and central England, but not common.
    Known at a few localities in Ireland and Wales. Some
    immigration from the continent.
  striolatum (Charpentier, 1840), Common darter: com-
    mon throughout much of England, Wales and Ire-
    land. Immigrates from the continent.
  vulgatum (Linnaeus, 1758), Vagrant darter: no recent
    records of this species. A rare immigrant.

Leucorrhinia Brittinger, 1850
    dubia (Vander Linden, 1825), White-faced darter: scarce, but found at a few scattered localities in England and in Scotland.

Appendix 2. The times of appearance of adult British dragonflies

The times of appearance of adult British dragonflies are shown in approximate seasonal order. The broken lines indicate times when very young or very old adults may be found. In exceptional seasons a few adults may be found earlier or later than the times indicated. (After Hammond, 1983.)

Pyrrhosoma nymphula
Brachytron pratense
Libellula depressa
Calopteryx virgo
Coenagrion armatum
Coenagrion puella
Cordulia aenea
Enallagma cyathigerum
Erythromma najas
Gomphus vulgatissimus
Ischnura elegans
Libellula quadrimaculata
Orthetrum cancellatum
Coenagrion pulchellum
Leucorrhinia dubia
Libellula fulva
Aeshna isosceles
Calopteryx splendens
Coenagrion lunulatum
Cordulegaster boltonii
Ischnura pumilio
Platycnemis pennipes
Somatochlora arctica
Aeshna caerulea
Anax imperator
Coenagrion hastulatum
Coenagrion mercuriale
Orthetrum coerulescens
Oxygastra curtisii
Ceriagrion tenellum
Coenagrion scitulum
Lestes dryas
Somatochlora metallica
Sympetrum fonscolombii
Sympetrum nigrescens
Sympetrum sanguineum
Aeshna juncea
Lestes sponsa
Sympetrum vulgatum
Aeshna cyanea
Aeshna grandis
Sympetrum danae
Sympetrum flaveolum
Sympetrum striolatum
Aeshna mixta

| MAY | JUNE | JULY | AUGUST | SEPT | OCT |

Index

Aeshna affinis, 46
A. caerulea, 4, 8, 52, 62, 63
A. cyanea, 8, 9, 26, 37, 42, 47, 51, 62, 63
A. grandis, 4, 8, 15, 41, 52, 62
A. isosceles, 4, 51, 62, 63
A. juncea, 7, 8, 19, 20, 52, 62, 63
A. mixta, 4, 8, 9, 15, 52, 62, 63
age of dragonflies, 30
Anax imperator, 4, 7, 9, 10, 14, 15, 37, 51, 62
A. junius, 23
A. parthenope, 46
A. tristis, 7
Anisoptera, 1
Anisozygoptera, 2
antennae, 24
auricles, 33

Brachytron pratense, 4, 22, 51, 62
British Dragonfly Society, 5, 73
bursa, 35

Calopteryx maculata, 37, 42
C. splendens, 8, 10, 17, 18, 38, 47, 55, 58
C. virgo, 4, 8, 29, 47, 55, 58
caudal lamellae, 49, 50
Ceriagrion tenellum, 4, 8, 56, 57
claspers, 33
Coenagrion armatum, 3, 60
C. hastulatum, 4, 8, 56, 60
C. lindeni, 46
C. lunulatum, 4
C. mercuriale, 4, 8, 50, 56, 61
C. puella, 7, 22, 56, 61
C. pulchellum, 4, 56, 61
colour, 32
conservation, 68
copulation, 42
  stages of, 43, 44
Cordulegaster boltonii, 8, 9, 13, 32, 33, 37, 47, 51, 62
Cordulia aenea, 4, 10, 32, 39, 53, 64, 65
County naturalists' trusts, 68
courtship, 41
Crocothemis erythraea, 42, 44

dissection, 71

economic importance of dragonflies, 3
emergence of adults, 14
  mortality during, 14
  time of, 14
Enallagma cyathigerum, 4, 8, 9, 22,
24, 25, 29, 30, 40, 43, 44, 46, 47, 48, 50, 56, 60, 61
E. hageni, 30
endangered species in Britain, 4
Erythromma najas, 8, 46, 49, 55, 58
extinct species in Britain, 4
exuviae, 15
eyes, 24, 25

feeding
  patrolling during, 27
  perching during, 27
female choice, 38
flight
  form of wing beats, 17, 18
  metabolic rates, 19
  oxygen supply, 19, 20
  speed, 118
  tethered, 23
  wingbeat frequency, 18
  wingloading in, 18
food, types of, 28

Gomphus vulgatissimus, 15, 50, 64
genitalia
  evolution of, 33
  secondary, 33, 34
giant dragonflies, 1
guarding, 46

habitats
  creating new, 68, 69
  destruction of, 1, 9
hamules, 34
Hemianax ephippiger, 30, 46
hibernation, 29
hovering, 19

Ischnura elegans, 7, 9, 10, 28, 29, 40, 41, 42, 43, 45, 56, 59
I. pumilio, 8, 56, 57, 59

larvae
  arrested development in, 7
  distribution of, 8
  duration of development, 7
  feeding in, 9
  habitats of, 8
  jet propulsion, 12
  labial mask, 9
  number of moults in, 7
  prey detection by, 9
  preyed on by, 8
  recording ventilation, 13
  respiration by gills in, 12
  salinity tolerance of, 9
  territorial behaviour in, 11

84   *Index*

Lestes dryas, 4, 55, 58
L. sponsa, 4, 15, 46, 55, 58
Leucorrhinia dubia, 8, 39, 54, 64, 65
Libellula depressa, 6, 8, 32, 54, 64, 65
L. fulva, 4, 8, 54, 64, 65
L. quadrimaculata, 4, 8, 22, 31, 42, 54, 65
ligula, 34
locality guides, 5
longevity, 29, 30

marking
  adults, 70
  larvae, 71
maturation, 29
Meganuridae, 1
micropyle, 6
migration, 30
monocular, field, 70

nodus, 1

ocelli, 25
ommatidia, 25
Onychogomphus forcipatus, 32
Orthetrum brunneum, 22
O. cancellatum, 8, 22, 36, 40, 42, 53, 64, 66
O. coerulescens, 8, 22, 36, 37, 53, 64, 66
oviposition
  by captured females, 48
  endophytic, 47
  exophytic, 47
  submerged, 46, 47
ovipositor, 35
Oxygastra curtisii, 3, 53, 64, 68

Palaeoptera, 1
Pantala flavescens, 18, 30
parasitism of eggs, 6
patrolling, 37
penis, 34
penis flagellum, 43
Petaluridae, 2
photography, 70
Plathemis lydia, 38
Platycnemis pennipes, 8, 55, 57, 60
preserving dragonflies, 71
predator avoidance, 31, 32
pseudopupil, 25
pterostigma, 1, 17

Pyrrhosoma nymphula, 8, 10, 14, 50, 56, 57

recognition of species and sex, 39
recording scheme, 69

satellites, 38
sexual dimorphism, 29, 32
signals, 40, 41
Somatochlora arctica, 4, 53, 64, 65
S. metallica, 53, 64, 65
sperm
  stored in males, 34
  storage in females, 35
sperm competition, 42
spermatheca, 35
spring species, 15
summer species, 15
Sympecma, 29
Sympetrum danae, 8, 54, 64, 66
S. depressiusculum, 22, 40, 46
S. flaveolum, 3, 66
S. fonscolombei, 3, 66
S. nigrescens, 54, 66
S. sanguineum, 4, 47, 54, 65, 67
S. striolatum, 4, 8, 24, 30, 54, 60
S. vulgatum, 3, 66

tandem, 40
  mistakes in formation of, 40
temperature control, 20
temperature recording, 23
territorial behaviour, 35, 36

ultraviolet light, sensitivity to, 26

vision
  colour, 25
  parallax in, 26
  polarised light detection, 26
  stereoscopic, 26
visual acuity, 25
visual sensitivity, 27
visual tracking, 26

wanderers, 38
warming up, 20, 22
wing structure, 16
wing-whirring, 22, 45
winter emergence of adults, 7

Zygoptera, 2